DREAM REAPER

A NOTE ABOUT THE AUTHOR

Craig Canine grew up in Des Moines, Iowa. He earned degrees in literature from Princeton University and New College, Oxford, which he attended as a Rhodes Scholar. He has worked as an editor for *Newsweek* magazine, Rodale Press, *Harrowsmith Country Life,* and Meredith Corporation. His writing has appeared in *The Atlantic Monthly, Eating Well, Reader's Digest,* and other national magazines. He is now an independent writer living in Palo Alto, California, with his wife, two children, and a menagerie of pets. *Dream Reaper* is his first book.

DREAM REAPER

The Story of an Old-Fashioned Inventor in the High-Tech, High-Stakes World of Modern Agriculture

CRAIG CANINE

The University of Chicago Press

Published by arrangement with Alfred A. Knopf, Inc.
The University of Chicago Press, Chicago 60637

University of Chicago Press Edition 1997
Printed in the United States of America

02 01 00 99 98 97 6 5 4 3 2 1

Grateful acknowledgment is made to Warner Bros. Publications Inc. for permission to reprint an excerpt from "I Cross My Heart" by Steve Dorff and Eric Kaz, copyright © 1992 by Warner-Tamerlane Publishing Corp. (BMI), Dorff Songs (BMI), and Zena Music (ASCAP). All rights o/b/o Dorff Songs administered by Warner-Tamerlane Publishing Corp. (BMI). All rights reserved. Used by permission of Warner Bros. Publications Inc., Miami, Florida 33014.

Library of Congress Cataloging-in-Publication Data

Canine, Craig.
Dream reaper : the story of an old-fashioned inventor in the high-tech, high-stakes world of modern agriculture / Craig Canine.
p. cm.
Originally published: New York : Knopf, 1995.
Includes bibliographical references and index.
ISBN 0-226-09265-8 (paper : alk. paper)
1. Harvesting machinery—United States—History. 2. Agriculture—United States—History. 3. Underwood, Mark. 4. Lagergren, Ralph. I. Title.
TJ1485.C36 1997
338.1′61′0973—dc21 96-49988
CIP

∞ The paper used in this publication meets the minimum requirements of the American National Standard for Information Sciences—Permanence of Paper for Printed Library Materials, ANSI Z39.48-1984.

To Molly

and to the memory of my grandparents

Elvin Fletcher Canine

(1888–1965)

Clara Elizabeth Perkins Canine

(1898–1988)

Ira Joseph Melaas

(1897–1980)

Madelia Florence Kalsem Melaas

(1898–1936)

If you eat, you are involved in agriculture.

—*1970s bumper sticker*

FIRST BOOKS IN THE SLOAN TECHNOLOGY SERIES

Dream Reaper by Craig Canine

Dark Sun: The Making of the Hydrogen Bomb by Richard Rhodes

Turbulent Skies: Commercial Aviation in the Twentieth Century
by Thomas A. Heppenheimer

PREFACE TO THE SLOAN TECHNOLOGY SERIES

TECHNOLOGY IS THE application of science, engineering, and industrial organization to create a human-built world. It has led, in developed nations, to a standard of living inconceivable a hundred years ago. The process, however, is not free of stress; by its very nature, technology brings change in society and undermines convention. It affects virtually every aspect of human endeavor: private and public institutions, economic systems, communications networks, political structures, international affiliations, the organization of societies, and the condition of human lives. The effects are not one-way; just as technology changes society, so too do societal structures, attitudes, and mores affect technology. But perhaps because technology is so rapidly and completely assimilated, the profound interplay of technology and other social endeavors in modern history has not been sufficiently recognized.

The Sloan Foundation has had a long-standing interest in deepening public understanding about modern technology, its origins, and its impact on our lives. The Sloan Technology Series, of which the present volume is a part, seeks to present to the general reader the story of the development of critical twentieth-century technologies. The aim of the series is to convey both the technical and human dimensions of the subject: the invention and effort entailed in devising the technologies and the comforts and stresses they have introduced into contemporary life. As the century draws to an end, it is hoped that the Series will disclose a past that might provide perspective on the present and inform the future.

The Foundation has been guided in its development of the Sloan Technology Series by a distinguished advisory committee. We express deep gratitude to John Armstrong, S. Michael Bessie, Samuel Y. Gibbon, Thomas P. Hughes, Victor McElheny, Robert K. Merton, Elting E. Morison, and Richard Rhodes. The Foundation has been represented on the committee by Ralph E. Gomory, Arthur L. Singer, Jr., Hirsh G. Cohen, Raphael G. Kasper, and A. Frank Mayadas.

Alfred P. Sloan Foundation

CONTENTS

DREAM REAPER

THE DREAM

A few can touch the magic string
And noisy Fame is proud to win them:—
Alas for those that never sing,
But die with all their music in them!

—Oliver Wendell Holmes, "The Voiceless"

The reaper was America's answer to Malthus.

—Herbert Casson, *The Romance of the Reaper* (1908)

THE FIRST TIME I saw Mark Underwood only his legs were visible. The rest of him was inside a combine, hidden in a thicket of belts, pulleys, armatures, axles, hydraulic lines, and steel struts. This, I soon learned, was a characteristic pose: Mark within the bowels of his mechanical leviathan, improving its digestion. He spent so much time in there that he might have set up housekeeping and lived for a while, like Jonah or Geppetto.

I shuffled my feet in the gravel so Mark would know someone was outside, then called out, "Hello in there!"

He wriggled out of the machine, wiped a hand on his blue jeans, and held it out to shake. Besides the jeans, he wore a black T-shirt smudged with dirt, white sneakers, a green seed cap, and metal-rimmed glasses. He stood a bit under six feet tall, with a frame lithe enough to squeeze into the combine's deepest recesses yet husky enough to make him look like a champion arm-wrestler. He looked, in sum, like the ordinary thirty-nine-year-old Kansas dirt farmer that, in one respect at least, he was. Less obviously, Mark Underwood was also an extremely talented inventor. Though he had struggled to finish high school and still found reading a chore, some people called him a mechanical genius, an Edison of agriculture. Mark preferred the title "ungineer."

The ungineer's main accomplishment at the time of our first meeting was the combine that had, moments before, seemed to swallow his upper body. In size and shape, it was a standard combine—the machine farmers use to harvest wheat, corn, rice, and the other grains that feed the world. Any farmer could tell you that Mark's machine was an International Harvester Axial-Flow model. Any farmer would also remark that such combines are normally painted red, and that Mark's looked like a strange albino, painted white from top to bottom. Even more unusual than its color were the murals painted on its sides. On one sheet-metal surface was a picture of a farmer wearing overalls and a straw hat, picking corn by hand. The tableau on the other side depicted a 1930s-vintage combine harvesting wheat, with a black dog running alongside it. On the rear end of the machine, painted in black script, were the words: "Dedicated to our parents, who dared to let us dream."

"Us" referred to Mark and his cousin Ralph Lagergren. Ralph was Mark's partner in the combine project—a salesman and marketing specialist who took care of fund-raising and strategy while Mark concentrated on inventing. On the day I arrived, Ralph was out visiting prospective investors; Mark expected him back that evening.

The unusual exterior of the cousins' combine only hinted at the even more radical changes that Mark had made inside. He had originally bought the machine, used but unaltered, for roughly twice what my parents once paid for a split-level house in the suburbs (or about $60,000). Then he attacked the combine with a cutting torch. He eviscerated it—pulped it like a dissected frog. Finally he inserted new guts into the shell, guts of a radical new design that he had invented and built himself. He called his invention the Bi-Rotor combine. The retrofitted International standing next to us was Mark's first full-size prototype. Its nickname was Whitey.

"A combine performs three operations," Mark explained. "It cuts the grain, then threshes it—removes the kernels from the ear—then separates the grain from the chaff. Standard combines do this with a cylinder that rotates inside a stationary, concave grate. Rasp bars on the cylinder rub the ears against the concaves, which knocks off the kernels. Then the kernels fall down through holes in the grates.

"This same basic system has been used for the past hundred years. It works, obviously, but after a century of refinement it's still not very efficient. Only seventy to eighty percent of the grain is knocked out in the threshing chamber. It takes the whole rest of the combine—another ten feet of complex machinery—to recover the rest of the grain. And even then, a lot of it winds up getting dumped out the rear end of the machine, along with the chaff.

"My approach is completely different." He beckoned me to join him under one of Whitey's mural-adorned side panels. The panel was propped open like the wing of a giant chicken gathering its man-size brood. "Instead of having a rotor turning inside a stationary concave, I've got an active grate," he said, pointing deep into the heart of the machine. "It's a perforated cage that turns around the outside of the rotor. It spins in the same direction as the rotor, but at a slower speed. Threshing takes place over three hundred and sixty degrees, all around the rotor, not just over a little section at the bottom." From where I stood, I could see part of the cage. It looked like a cylindrical

Whitey in a wheat field. (Mark Underwood and Ralph Lagergren)

colander, perforated with holes the size of the end of my little finger. "These holes are smaller than the ones in a standard concave grate, so less junk gets through," Mark continued. "I accomplish both the threshing and the separation in one step, like *that*"—he snapped his fingers. "And all within a distance of four feet. You can eliminate at least half the moving parts of the combine. All of that mess back there." He waved at the machine's elongated abdomen. "With the Bi-Rotor design, you can do a better job with a much smaller, less complicated machine."

The white combine was parked in front of the machine shop on Mark's farm, about five miles from the village of Burr Oak, in north-central Kansas. Mark's closest neighbors were relatives. When a pickup truck passed on the gravel road in front of Mark's house, raising a cloud of brown dust, the chances were better than 50 percent that its driver was either his father, one of his two brothers, or his crazy cousin Dewey. All of them farmed on and around the ground that Mark's grandfather had broken out of prairie sod in the 1880s. Mark, his wife, and their three children lived in the farmhouse that his grandfather had built. Like many other farmhouses in the area, the Underwood place was painted yellow with brown trim.

In October, the time of my visit, yellow and brown were also the colors the earth wore. The land surrounding Mark's farm spread out like a rumpled quilt—a rolling patchwork of fields, with a few trees growing along the low creases. Scattered here and there in the quilt of autumn hues were several patches of brilliant green. These were fields of winter wheat. The wheat had been planted only a month before, and now stood about three inches tall. Its grassy spears would stand dormant through the winter, then shoot up in the spring, turning a premature autumn-gold in July.

In the fall of 1991 it was a wonder that the wheat had come up at all. Much of Kansas was baking through its third straight year of drought. "Down here," Mark said as he pointed to a field of parched corn, "seventy or eighty bushels of corn to the acre is considered good. This year it'll be less than that. We've had ten inches of rain all year. That's desert conditions. I don't know how we made a crop at all on that."

We walked to a Winnebago camping trailer parked on the Underwoods' brown lawn, between the road and the house. The camper served as Mark's mobile office, where he did most of his drawing and patent-application work. He either held or had applied for five patents.

Three of them dealt with various aspects of the Bi-Rotor threshing mechanism, and the fourth covered an expandable grain head for a combine. "The fifth patent I can't talk too much about right now," he said, "since we haven't filed the final application. But it's another major combine innovation."

Inside the trailer, he pulled out a few technical drawings of the Bi-Rotor for my inspection. But what he really wanted to show me was his well-thumbed copy of a book called *The Grain Harvesters,* a history of harvesting technology from the ancient Sumerians to the present. "Look at this," Mark said eagerly, opening the book. He pointed to a paragraph about the role that inventive farmers such as himself had played in the development of the modern combine. He had underlined the following sentence: *"Farmer-innovators have made and will continue to make major contributions to the industry and some of the leaders and innovators of modern industry were themselves farmers—McCormick, Harris, Massey, Baldwin."* Next to this list of distinguished farmer-inventors, Mark had added his own initials, along with "BI-ROTOR — 1982." He grinned, a little sheepishly.

"So you've been working at this idea for almost ten years?" I asked.

"Yup. And now that we've got ol' Whitey here, it's just starting to get interesting. We just finished building Whitey last spring. This is her first harvest season. So far we've cut wheat, milo, soybeans, and a little corn with her, and she's worked great. But Whitey's just the first prototype. This winter we'll start building a second prototype—a totally new combine from the ground up. The next one will incorporate *all* of my innovations. It will be the fulfillment of these words. The Bi-Rotor will be the next chapter in the history of grain-harvesting technology." It would be Mark Underwood's ultimate dream reaper.

I HAD ARRIVED at the Underwood farm in midmorning, just as Mark was preparing to take Whitey out to harvest corn. I was about to witness the Bi-Rotor's first sustained tests in America's major crop. Mark enlisted me to help him ferry machinery—the combine, a grain truck, Mark's red pickup—to a cornfield a mile from his house. Once everything was ready, he sprang into Whitey's cab and gunned the engine to full throttle. He maneuvered the big machine toward the brittle cornstalks, then attacked them with a vengeance.

The normal operating speed for a combine is about three miles an hour, but Mark was going a good deal faster than that. He pushed the machine hard to see what it could do, just as he once tested the limits of his souped-up cars on the country roads around Burr Oak, inspiring a kind of horrified admiration in his high-school classmates.

Mark had made several passes through the corn when a pickup drove into the field and a man got out. The man introduced himself to me as Von Van Meter, Mark's partner in a small bulldozing business. Von, who was built like the bulldozers he operated, walked over to the corn stubble that Mark had just mowed down with the machine. He peered at the ground. With his toe, he probed around, looking for unthreshed cobs, loose kernels, any sign of lost grain. "Man, that's clean," he said. Then he looked up at the roaring combine, spat out a wad of chewing tobacco, and chuckled.

"He's going six, maybe six and a half, miles an hour," Von said. "He thinks he's driving a sports car. He's quite a guy—a different kind of a cat. Folks around here all know about him and his machine. They're rooting for him like he's a local sports hero. If he can just get one of the major farm-equipment companies to come up and take a good look at it."

Mark stopped at the end of a nearby row. He poked his head out of the door of the cab and yelled, "Seven miles an hour!"

After his next pass Mark drove close to where Von and I were standing. He stopped the machine and hopped out to check the grain bin. "I was flying!" he shouted. "And look, the field's clean."

Von poked the ground behind the combine some more with his foot. "Find a kernel," Mark challenged.

"Can't," said Von. "Boy, you were moving. It looked like you were in road gear."

"I'm going to take a few more flying swipes."

"Let 'er roll, boy!"

At lunchtime Mark and I bounced back to his house in the pickup. As we pulled into his driveway he said, "There are so many people out there with good ideas who never get the opportunity to follow up on them. Even if this combine idea never goes anywhere, at least I won't get to be eighty years old and say, 'You know, I had this idea once, but I never did anything with it.'"

Walking toward the house, I spied a flash of bright red in one of the outbuildings on the farm and asked Mark what it was.

"That's my pride and joy," he said. "It's a nineteen-sixty Ford Sunliner convertible that I restored. Someday when I'm halfway successful, I'm going to get me a cigar and smoke it while I'm driving around with the top down."

Mark's wife, Deborah, had fixed a big lunch—fried chicken, mashed potatoes and gravy, fresh tomato slices, deviled eggs, dinner rolls, boiled peas, iced tea, and, for dessert, cherry pie and a toothpick. She struck me as a matter-of-fact woman, solid in every sense of the word, with a dry sense of humor lurking beneath the surface. I asked her what she thought of this whole combine project.

Deb, as she's called, frowned as she considered the question. "Well, it means he's gone a lot," she said, nodding her head toward Mark. "Last year he was home a total of forty days. The rest of the time he was in Texas, working on that combine. But I keep busy." She told me that she was a leader of the 4-H Club in which her three children were active. She also served on church committees, ferried the kids around to their many sports events and 4-H meetings, and ran her own business preparing income-tax returns in the winter. The combine didn't interest her much. "We hope something comes of it," Deb said, "but I've never got my hopes up yet, or I could be hoping forever. It's been ten years now that he's worked on this idea."

Mark chimed in: "We're still in the embryo stage, trying to get born. We hope the umbilical cord doesn't get broken before we have a chance to get out."

The next morning Mark took Whitey back into the corn. "It's cutting kind of hard," he said after twenty minutes in the field. "I can tell by the sound of the corn going through the rotor that the corn is wet." To be accepted for storage at an elevator, corn must test at no more than fifteen percent moisture by weight. Any wetter than that and it will spoil in the silos, creating the threat of fire and explosion. Also, the elevator doesn't want to pay for water when what it wants is grain.

Mark decided to get a moisture reading on the corn at the nearest elevator. He needed a coffee can to hold the sample, so he went to fetch one at his parents' house, a quarter-mile down the road.

"Corn's wet," Mark announced as he entered the kitchen of the house in which he had grown up. His father, Delbert Underwood, sat at the kitchen table sipping coffee. At the age of eighty-three, he no longer farmed actively, but he kept track of the daily movements of his three sons with the alert interest of an air-traffic controller.

"How wet?" his father asked.

"Don't know. I'm going to take a sample uptown."

His father shook his head. "I don't see how the corn could be wet when the weather's so dry. I've seen it this dry before, but I never hope to see it any drier. It was this dry in the thirties, and again in the fifties. One more year like this and we'll be done for. In 1935 the dust in this house was this deep." He held his hand a few inches off the floor. To me, he added, "It's probably still that deep in the attic. My folks moved here from Ioway"—he used the old-fashioned pronunciation favored by many Kansans—"because they thought it was too wet and swampy up there. I guess they found a drier spot, all right." He smiled ruefully.

Mark found a coffee can and drove back to the combine, where he filled the can with corn from the hopper. Mark's father and a brother, Gene, followed us to the field in Delbert's pickup. Mark passed the can around. The three Underwoods each took a few kernels, popped them into their mouths, and chewed on the corn thoughtfully.

"Yeah, it's wet all right," Mark's brother said. "I'd say around nineteen percent."

"That's what I was a'thinkin'," said Mark. His father nodded.

On the way to the elevator Mark and I passed a dusty crossroads. "This used to be the town of Otego," Mark said. "My mom was born and raised here. It once had a population bigger than Burr Oak's. Here was the lumberyard. The church sat here, the bank was over there, and the schoolhouse was right there." Only trees and weeds now stood where he pointed. "Otego went belly-up in the dirty thirties, when a lot of the farmers quit and moved out. A few families still lived here when I was a kid. Now there's nothing. I've seen a lot of farmers disappear off the scene in my time, and I'll see more."

I asked how farming had been treating him.

"Well, I've been tapering off," he said. "Three years ago I was farming a thousand acres. Now I'm down to around six hundred. I don't have much time for it now, between the combine project and excavation work. I'm not too sorry to be farming less, though. Farming's no way to try to make a living."

Mark's mood suddenly grew intense, his brow furrowed. "You know, sixty-five percent of the farmers today are over sixty-five years old," he said. "They're a dying breed. It's sad. We've kind of got our priorities mixed up in this country. We spend thirty-five percent of our income on recreation and seventeen percent on food. Those figures should be flip-flopped. Food's too cheap. We're getting nineteen-sixties prices for our crops. My dad got two dollars and thirty cents a bushel for corn in the sixties. That's about what the elevator will pay me today. But our expenses are *way* up. A young feller can't get started in farming because the costs are so high. A new tractor costs seventy-five thousand dollars. A combine costs a hundred and thirty thousand. A decent pickup costs twenty thousand. Then a drought like this comes along and we can't get a decent crop *or* a decent price. We can't hardly make it out here.

"My machine will be a lot more efficient than what's out there now," he added, his mood brightening a little. "That means a farmer will have less cost per acre wrapped up in harvesting. You know the picture of the old guy picking corn on Whitey? A lot of people say he looks like Ronald Reagan. I say we're trying to practice a little Reaganomics here—get some money to trickle down to farmers with a more efficient combine."

A problem in logic lurked here, one that reached beyond Mark and his machine. The same quandary made modern agriculture simultaneously one of the greatest wonders and saddest tragedies of Western civilization. I framed the problem to Mark. "Isn't the invention of more and more efficient machines, like yours," I said, "part of the reason why grain prices are so low, and why so many farmers have had to leave their farms? I mean, American agriculture is already plagued with overproduction. By making it even easier for farmers to produce more grain, won't the Bi-Rotor just make a few more family farms unnecessary?"

A wounded look passed across Mark's face. "I guess that's what a lot of people strive for—to create something new that will put them on the other side of the fence," he said, a little sharply.

Then he reflected for a moment. In that moment I believe he saw the contradiction I had raised and decided to accept it. As a farmer, Mark could see how changing technology contributed to the forces that had wiped out Otego—the same forces that continued to drive farmers off the land. As an inventor, however, he gave little thought to such matters. He furnished living proof that the wellsprings of innovation flow regardless of the consequences, and from no baser motive than the desire to create something new, and maybe even start a retirement account with the proceeds.

THOMAS JEFFERSON, the founding father of American agrarianism, embodied this contradiction himself and ingrained it in our national psyche. Because of Jefferson we worship the farmer as well as the scientism that has made farmers nearly extinct. Jefferson himself never had to confront the contradiction, but our land has been forever transformed by it.

Jefferson the agrarian idealist believed fervently in the idea that democracy could flourish only in a nation where self-sufficient farmers made up the majority of the population. "Cultivators of the earth," he wrote in one characteristic passage, "are the most valuable citizens. They are the most vigorous, the most independent, the most virtuous, and they are tied to their country, and wedded to its liberty and interests, by the most lasting bonds."

But Jefferson the agrarian was also Jefferson the technological enthusiast, founder of the patent system. Though he firmly believed in a republic of farmers, he also held an Enlightenment faith in science and technical progress. He was always among the first to hail a new labor-saving device, especially if it had to do with agriculture. The gentleman farmer of Monticello invented a threshing machine, a device for extracting the fibers from hemp stalks, a mechanical seeder, and an improved moldboard plow. "Science never appears so beautiful," he once wrote to an English acquaintance, "nor any use of it so engaging as those of agriculture and domestic economy."

The idea that farmers could keep up with, let alone surpass, the demand for food and fiber in America's burgeoning cities was unimaginable in Jefferson's day. Hunger and scarcity were daily facts of life, threatening peace and political stability, raising the specter of revolution across Europe. Jefferson himself was in Paris as American envoy to France when the French Revolution broke out in 1789. Although he dismissed the idea of food shortages as a contributing cause of the French turmoil, he helped fund soup kitchens for starving peasants.

In 1798 the British philosopher Thomas Malthus published his deeply pessimistic theory that population always grows geometrically, while "subsistence," or food production, increases only arithmetically. Jefferson read Malthus eagerly and found the 1803 edition of his *Essay on the Principle of Population* "masterly." Yet Jefferson did not consider it applicable to America—perhaps because, as the young country's president, he had just acquired the Louisiana territory. The United States, it seemed, had a surplus of arable land for the foreseeable future. The spectre of scarcity would cease to haunt its people. On the other hand, the notion of producing *too much* food remained inconceivable, beyond imagining.

So it came as a shock when, scarcely a century after Jefferson's death, American farmers experienced a sudden and serious decline in their prosperity. It was caused mainly by overproduction. Somehow food output had surpassed population growth, even as the ratio of farmers to urbanites steadily diminished. The story of how Malthus was proved wrong during that century—roughly the 1830s to the 1940s—is the story of modern agriculture, a story in which Mark Underwood and his invention play an emblematic role.

WHEN WE ARRIVED in the elevator office the man at the counter reached into Mark's coffee can, grabbed several corn kernels, and rubbed them together in his hands. "It *feels* wet," he said. The phone rang and the man went to answer it. He was gone for several minutes. When he came back Mark said, "The corn just dried a percentage point while you were talking on the phone." The man put a sample from the can into an electronic moisture tester, a device the size of a bread box,

The cousins in 1964: "We'd share our dreams, but not our soda pop."
(Mark Underwood and Ralph Lagergren)

that sat on the counter. A moment later the digital readout lit up: 19.8 percent moisture.

"That's kind of what I thought," Mark said, and we left.

Ralph Lagergren, Mark's salesman-cousin, arrived late that night in his cherry-red GMC pickup. Ralph had grown up in Lincoln, Kansas, two counties south of Burr Oak. The two cousins had seen little of each other during school years but often spent summers together on the Underwood farm, swinging on ropes in the haymow and setting off firecrackers. After high school, their paths diverged. Mark enrolled in a community college in nearby Beloit, while Ralph went off to Kansas State University in Manhattan. Diploma in hand, Mark returned home to the farm; Ralph, meanwhile, embarked on a career in sales and marketing that took him all over the country. He had lived most recently in Fort Worth, which accounted for the Texas plates on his truck. But now he and Mark were together again, their divergent training and talents complementing each other as they pursued a common dream. The dream, in essence, was to leave their mark on the world, just as they had left their initials scratched in tree bark and fence posts when they were young.

Ralph had just driven about 500 miles in order to meet, early the next morning, with some important visitors who wanted to see Whitey. The visitors were from Case-International Harvester, the nation's second-largest farm-equipment manufacturer (whose hyphenated name represented the combination of International Harvester and J. I. Case, two old farm-machinery companies that were both acquired in the 1980s by Tenneco, the petroleum and manufacturing conglomerate). Ralph's main contact at Case-IH had tried to make his interest in the Bi-Rotor sound as casual as possible. He and a few other combine specialists, he said, planned to be in the Burr Oak area on other business and wanted to stop by to see the machine. Ralph couldn't imagine what "other business" they might have. Burr Oak lay approximately fifteen miles from the geographic center of the lower forty-eight United States, which is to say, the middle of practically nowhere. "The closest airport is eighty miles away," Ralph told me. "There isn't a lot else going on within a hundred-mile radius. To get to this part of Kansas, you have to *want* to come here."

After his morning with the delegation from Case-IH, Ralph knocked on the door of my room at the Dreamliner—one of those U-shaped, one-story motels that sprang up along every two-lane highway in America during the 1950s, before interstates siphoned away all the long-distance traffic. The Dreamliner, with its permanently drained swimming pool, had seen better days.

"They were pretty impressed," Ralph said of the corporate scouts. "They watched Mark take some flying passes through the corn. They couldn't believe how fast he was operating that combine. When they looked behind the machine for lost grain and couldn't find any, they were amazed. Of the five men who were here, I'd say three of them would go with the project right now. But they'd have to go to the corporate hierarchy and convince them to retool for a totally new combine. We haven't developed the machine yet to where they could see light at the end of the tunnel. But we will. We've got the ball, now we've got to run with it."

Ralph knew what it was like to "go up against the corporate hierarchy," as he put it. For more than fifteen years he had worked for various large corporations and had grown tired of fighting their bureaucracies. Most recently, he had put in eight years as a pharmaceutical salesman. "I was working for Johnson and Johnson," he said. "Good salary, great benefits, job security—the works. But every time I had an idea, I had to go ask somebody, and they'd try to shoot it down. After three or four years of that, I stopped asking. I started going about things my own way, not by the book."

I asked him what he meant.

"There's the time I threw a nurse-appreciation party," he recalled. "Nurses are key to the success or failure of a pharmaceutical salesman. They have the power to either smooth your way or block your access to the doctor. Nurses were always doing nice things for me, helping me out, so I decided I would throw a party for them. I asked my boss about it and she said, 'Oh, no. You're not going to do that.'

"So I decided to do it myself. I took a few days' vacation and went to work. I got Nieman Marcus in Dallas to donate a beauty makeover. I got tickets to Six Flags, gift certificates to McDonald's—a bunch of things like that. Then I got a friend of mine who does a great impersonation of Dana Carvey characters from *Saturday Night Live.* Another friend does an incredible Elvis. So we worked up this crazy skit. I was

Superman, who swooped in to plug Johnson and Johnson products. The Sheraton donated the use of its ballroom downtown. I had it all lined up.

"Then I went back to my boss. I told her what I'd done, and said that all the company would have to do was pay for some wine and cheese. She said, 'Well, I guess we could do that.'

"The party was a tremendous success. I'm certain that sales benefited. Salesmen from other companies started giving me shit. They told me the nurses would say to them, 'Ralph threw us a party. What have *you* done for us?' It was a good idea. But I had to arrange most of it on my vacation time."

Though Mark and Ralph share a set of grandparents, they don't appear to share many genes. In many ways they couldn't be more different. Mark is solid, round-featured, dark-haired, and sometimes brooding; Ralph is more angular, slighter of build, lighter in complexion, and always glows with enthusiasm. They had worked long days together, seven days a week for months on end, with a remarkable absence of friction. It helped, no doubt, that their responsibilities were clear-cut and didn't overlap. "Mark has a vision of how the machine works mechanically, and I have a vision of how to market it," Ralph told me at the Dreamliner. "Mark knew exactly how to modify an existing combine to build our first full-scale prototype; it was my idea to paint the machine white and put the murals on the sides. That won us a lot more attention than if it were just another red combine. Little marketing touches like that have been just as important to our success, so far, as the soundness of the mechanical ideas. It's been a pretty good partnership.

"I worked for farmers during wheat harvest when I was in high school, so I knew basically how combines worked and what the problem areas are. I understand enough about the mechanical side so I can play devil's advocate with Mark, to help him develop his ideas. But there aren't many people who can top his ability to work out solutions to mechanical problems."

Mark was one of Ralph's favorite topics of conversation. He talked about Mark with a mixture of boosterism and candor, as though his cousin were a new drug he was trying to sell.

"Mark's a terrific idea-man and inventor," Ralph said, "but he also has great shop skills. We built Whitey down in Texas last winter, and he had the professional welders in the machine shop down there admiring his welds. 'Man, he's good,' they'd say.

"He has an uncanny memory for visual details," Ralph continued. "When we were working on Whitey we needed a belt to turn the threshing cage. A D-belt, eighty-eight inches long. We went to a place that specialized in belts. They said you can't get a D-belt less than a hundred and ten inches long. Mark said he just knew he'd seen a D-belt shorter than that someplace. He thought maybe it was on an old Massey Ferguson machine that he'd seen a long time ago. There happened to be a Massey dealer up the road, so we drove over there, went to the parts desk, and asked to see the belts. Mark just walked right up to one and said, 'Here it is.' The parts guy took out a tape measure. It was eighty-eight inches long."

By this time Ralph and I had moved our conversation from the Dreamliner to a tavern in Mankato, the county seat near Burr Oak. The tavern was a big, open room with a few pool tables near the back. Up front, booths lined the wall opposite the bar. We sat in a booth while the jukebox played country-western tunes and some men wearing camouflage hunting caps and hooded sweatshirts played pool.

Ralph, I soon realized, *lived* to sell. When he wasn't selling Mark (or even when he was), he was selling himself, but so openly and with such good humor that I forgot I was in the presence of a salesman.

"Twelve years ago," Ralph said, "Mark told me this idea he had for a new kind of combine. I called him out of the clear blue sky two years later and said, 'Are you still interested in that idea?'

" 'You mean the combine?' he said.

" 'Well,' I told him, 'I'm coming home in two weeks. I want to see some drawings. If you can convince me that it's worth pursuing, I'll go to the end with it. If you don't convince me, I won't mention it again.'

Mark took the bait. He soon produced several drawings and built a little desktop model of his Bi-Rotor device. When he attached a hand drill to the miniature combine, it threshed wheat. Ralph was convinced. For eight years he wore out tire rubber and shoe leather in an effort to persuade investors to fund the development of the combine. "But," he recalled, "I ran into a brick wall. Nobody bit. I'd never had so much trouble selling anything in my life. Finally I asked myself: If I were an investor, would I give money to these guys—a wheat farmer and a drug salesman?"

Ralph realized that the time had come to either devote all his time and attention to the combine project or give it up. He knew immediately what he would do. The next day he gave notice at Johnson & Johnson. Two weeks later, he was on his own.

"I love quotes," he told me, "and there's this one from Oliver Wendell Holmes about how we're all born with a music box inside, but most people die with the music still in them. Well, I wasn't going to do that. Mark and I both knew that this was a chance to live our dream. Those chances don't come along very often. Quitting a good job is a scary thing to do, but I've never regretted it for a moment."

He leaned over the table and asked, "Do you know the true definition of an entrepreneur?"

"I guess not," I said.

"An entrepreneur is someone who's willing to step out and live in sheer terror every day. There's some excitement to that. Ever since I quit my job, I knew there was no turning back. It's kind of like when Cortès landed in Mexico. He burned his ships, and after that, his men were sufficiently motivated."

ROMANCE OF THE REAPER

This magical machinery of the wheatfield solves the mystery of prosperity. It explains the New Farmer and the miracles of scientific agriculture. It accounts for the growth of great cities with their steel mills and factories. And it makes clear how we in the United States have become the best fed nation in the world.

—Herbert Casson, *The Romance of the Reaper* (1908)

If the U.S. had to harvest its crops using the hand methods of the 1830s, then every American man, woman, and child would be needed in the fields every day during the harvest season.

—Graeme Quick and Wesley Buchele,
The Grain Harvesters (1978)

Horace Greeley, editor of the *New York Tribune* and one of the most influential Americans of his time, was touring the French countryside one summer when he spotted a band of peasants moving slowly through a field of ripened grain. They were stooped over double, laboriously harvesting the crop with short-handled sickles.

Greeley, who was deeply interested in agriculture, approached one of the peasants and asked, "Why don't you use a scythe? Then you could cut twice as much."

"Because," the Frenchman replied, "I haven't got twice as much to cut."

The scene Greeley witnessed took place in about A.D. 1850, though it might, with a few alterations of detail, have occurred in the Nile Val-

Scythers and rakers harvest wheat the old-fashioned way on the plains of Hungary.
*(*Wallaces Farmer *by Ewing Galloway, N.Y.)*

ley in 1850 B.C. or earlier. Tomb paintings and carved reliefs from ancient Egypt show slaves and peasants cutting grain with tools nearly identical to the ones Greeley saw slicing straw that day in France. From the age of the pharaohs to the age of Napoleon, methods of harvesting grain remained essentially unchanged. Even the hunter-gatherers of the Natufian culture, living on the Levantine lowlands around 9000 B.C., harvested wild grain with sickles made of bone and flint. Grain-cutting knives were straight at first; then, over the millennia, they gradually acquired angles and curves for greater efficiency. Reapers grasped a sickle in one hand and a bunch of grain in the other, sliced through the stems just below the grain heads, and laid the heads on the ground. A troupe of gatherers followed behind the reapers, carrying baskets. They filled the baskets and carried them to one end of the field. Finally, the piles of grain heads were carried to the threshing floor for the next stage of the long harvesting process. An experienced harvester with a razor-sharp sickle, working in thick grass, could cut perhaps half to three-quarters of an acre in a day. In an agrarian society—one, that is, in which the number of mouths to feed is approximately equal to the number of field workers—this pace is just about adequate to ensure subsistence.

A technological breakthrough came in the ninth century A.D., when some farmers in northern Europe modified the sickle with a longer handle and blade, thereby inventing the scythe. Nearly a thousand years later, in the 1790s, some American farmers began using lightweight scythes fitted with long projecting fingers called cradles. A cradle enabled the scytheman to deposit cut grain to one side of his swath in a neat windrow, making the stalks easier to collect. In a field of technology that had barely progressed since the Bronze Age, the cradle seemed revolutionary. A cradler and four or five additional laborers could cut and bundle into sheaves the grain from about three acres of good land per day.

This is about where things stood in the United States in 1831, the year Cyrus McCormick tested his first mechanical reaper.

Generations of schoolchildren have been taught that Cyrus McCormick invented the reaping machine. He did not. Nor was McCormick's reaper the first commercially successful machine of its kind. McCormick did, however, provide much of the creative energy and entrepreneurial drive that fueled the development and adoption

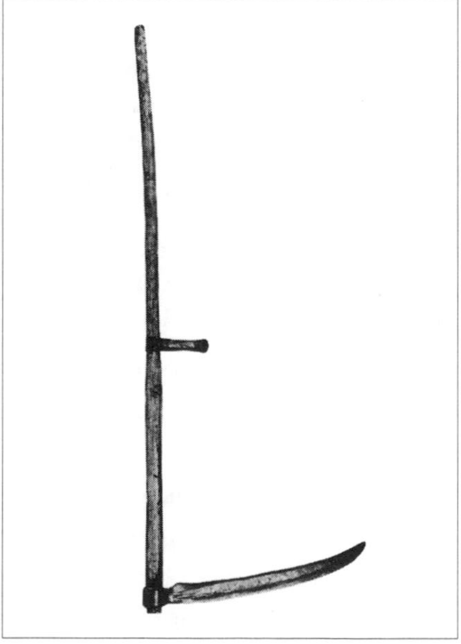

The progression from the sickle to the scythe represents thousands of years of harvesting technology. (Smithsonian Institution)

of the reaper and many other agricultural machines. Like most so-called inventors of major technologies, he was more accurately a catalyst for changes that were already in the wind.

When Cyrus McCormick was born in 1809, most farmers in the United States used tools and techniques that a medieval serf would have recognized. Nine out of ten Americans lived on a farm. By the time McCormick died, in 1884, farmers had become a minority of the population. Machines such as the reaper had become nearly universal across the grain-growing regions of the United States. In the space of a single human lifetime, America had evolved from a rural, agrarian nation into a largely urban, industrial one.

McCormick not only helped to bring about this transformation but became one of its most fitting symbols. Like America, he was born on the farm and moved to the city. He became a prominent member of the country's first generation of industrialists. The implements he made were among the first examples of modern factory production. And the industry he led was one of the early successes of American capitalism. Cyrus McCormick was a great inventor, a master salesman, a prophet of mass production, and a robber baron, all rolled into one.

A STRANGE NOISE prompted Miss Polly Carson to rush to a front window of her house to see what was going on. What she observed on that July day in 1831 was so unusual that she remembered it vividly more than fifty years later. At the time, as she gazed at the clattering contraption passing by on the road, she had no idea what it was. Pulled by a team of nervous horses, the wide, low-slung machine whirred, rattled, and clicked as though it were alive. A wooden platform extended far out to one side, with a triangular board attached to the end like an outrigger. The platform was covered with a piece of canvas that snapped in the breeze, adding to the skittishness of the horses. Two men held the animals tightly by their bridles so they wouldn't bolt and smash the fragile-looking machine to pieces.

Miss Polly recognized the men leading the horses as Old Joe Anderson and Anthony, slaves of Robert McCormick and his son, Cyrus. The McCormicks were prominent farmers in that part of the Valley of Virginia, a remote region of high, rocky ground tucked between the Blue Ridge and Allegheny ranges of central Appalachia. Robert McCormick was a prominent member of the local Scots-Irish farming community. He had a reputation for being a rather contrary and opinionated man (to his neighbors' consternation, for example, he openly questioned Calvin's doctrine of predestination) and was also known as an obsessive tinkerer. He was *always* inventing things: a newfangled blacksmith's bellows, a gristmill, and (according to one of his several patents) "an improvement in teaching the art of performing on the violin." He had tried repeatedly to make a machine that would reap grain but had met only with failure. Just a month before Polly Carson heard the strange racket outside her window, Robert McCormick had tested his latest reaper at Walnut Grove, the family's farming estate. The results had been discouraging.

Now, however, the man who was shepherding the odd contraption past Miss Carson's house was not Robert but McCormick's twenty-two-year-old son, Cyrus. The young McCormick was taking his machine to a field owned by John Steele, the proprietor of a nearby tavern, to cut oats. Miss Carson did not follow the procession to see what happened next, but a neighbor who did watch told her that it was "a right smart curious sort of thing." Many observers probably thought Cyrus's machine was just another idle folly from the workshop of the excessively inquisitive McCormick family. Indeed, as Polly Carson later recalled, "nobody ever believed it would come to much."

Nevertheless, with his machine Cyrus McCormick cut six acres of John Steele's oats that day. McCormick was the first to admit that it wasn't a very pretty job. The field was left looking like the scalp of a towheaded boy with a bad haircut. Some of the oat heads were trampled into the ground, uncut; others were left standing in clumps with the grain intact; still others had been shaken so hard by the machine that the kernels had popped out of the heads and fallen to the dirt, where they were no good for anything but a foraging pig.

Still, McCormick was not entirely discouraged. A fair percentage of the oats in Steele's field had behaved the way he wanted them to. A long, oscillating knife blade at the front of the platform on his machine cut the stalks a few inches above the ground. The severed oats fell backward onto the canvas-covered platform. When a sheaf-size bunch, called a gavel, fell onto the canvas, someone walking beside the machine (most likely Joe or Anthony) took a wooden rake and swept it off the platform. Later a crew of binders would pick up the loose gavels and wrap them with a piece of straw, making sheaves that they would stack in shocks to dry.

With a McCormick reaper, two laborers and a horse could do the work of five expert cradlers. (International Harvester Collection, Wisconsin Historical Society)

Joe Anderson, who had known the young McCormick since he was a baby, was one of the few confirmed eyewitnesses of the scene in Steele's oat field that day. Old Joe summed up the event in biblical terms. "Old master Robert gave up working on the reaper when master Cyrus said he thought it could be done," Joe said. "It is like the good Lord who sent his son to save sinners. He began the work, but his son did the work and finished it."

The harvest of small grains was nearly over for the season by the time Cyrus McCormick tried his invention in John Steele's oat field. McCormick could do little more with the machine that year, so he traveled to the hemp-growing parts of Kentucky to peddle a hemp brake (a machine for separating the fibers from the stalks) that his father had invented. He found few takers. The next summer, back home in Virginia, he modified his reaper in hopes of making it work better, which it did. He cut fifty acres of wheat and oats with his machine at Walnut Grove, which gave him enough confidence to risk another public demonstration.

The trial was held on the farm of John Ruff, near Lexington, about eighteen miles from Walnut Grove. At least seventy-five people, including several of the most prominent farmers in Rockbridge County, gathered to watch. Cutting conditions were less than ideal in Ruff's field, which its owner's name homophonically described. The wheat was uneven; some of it had been beaten down by the wind and lay in a tangle just off the ground. McCormick's reaper performed abysmally. The inventor trudged alongside his machine in stony-faced embarrassment, making adjustments and minor repairs in a desperate effort to redeem the moment.

Some of his difficulties came from one of the modifications he had recently made. To prevent the reaper from flattening grain underneath it as it moved forward, McCormick had added a revolving reel. The reel looked like a flimsy paddle wheel turning just above the cutter bar. A pulley and belt, actuated by the ground-driven main wheel, caused the reel to rotate, pushing the grain against the cutter. A set of "fingers" projecting in front of the serrated cutter bar kept the stalks from moving from side to side as the reciprocating knife sliced through them. In theory the system was sound, but the speed of the reel proved critical. If it rotated too slowly, it failed in its function; if it went around too fast, its paddles flailed the grain, beating out the kernels. That was what it did on Ruff's field.

Farmer Ruff finally told McCormick that he wished his grain to be reaped, not threshed where it stood. He ordered the machine, the inventor, and the spectators off his field.

The demonstration would have ended ignominiously right then if William Taylor had not ridden up on horseback. Taylor owned the farm next to Ruff's. He told some bystanders to pull down a section of fence and invited McCormick to try the machine in his wheat. Taylor's field was smoother than Ruff's, and less of the grain had blown over. McCormick figured he didn't have much to lose, so he crossed over onto Taylor's land. Things went better this time. The reaper reaped.

Some accounts report that a certain Professor Bradshaw, who taught at a girls' academy in Lexington, watched the proceedings in Taylor's field for a while, then marched up to McCormick and announced solemnly, "That . . . machine . . . is . . . worth . . . a hundred . . . thousand . . . dollars." McCormick replied that he would gladly sell all rights to it for half that amount. No one stepped forward with an offer.

By the end of that harvest season, however, McCormick had raised his estimation of the reaper's value. He allowed himself the distant hope, the all but insanely optimistic dream, that someday his reaping machine might earn him *a million dollars.* "The thought was so enormous," he said later, "that it seemed like a dream—like dwelling in the clouds—so remote, so unattainable, so exalted, so visionary."

. . .

ABOUT NINETEEN CENTURIES before Cyrus McCormick, Pliny the Elder described the reaping machines of ancient Gaul. "On the vast estates in the provinces of Gaul," he wrote in his *Natural History,* "large frames [*valli*] fitted with teeth at the edge and carried on two wheels are driven through the grain by a pack animal pushing from behind; the ears thus torn off fall into the frame." The Gallic *vallus* is the first known device to harness animal power for harvesting grain.

Some 1,700 years after Pliny wrote this description of the vallus, it resurfaced in a Scottish agricultural journal. In 1785, after reading a translation of Pliny the journal published that year, one William Pitt of Stafford, England, wrote to the editor to suggest an improved vallus with a ground-driven "rippling," or reaping, cylinder. The rotating cylinder, mounted transversely on the front of a chariotlike cart, would snatch ears of grain from their stalks and throw them into the chariot's "great car." This chariot would be pushed from behind (like the vallus) by a horse. Pitt included a careful sketch with his letter. The Scottish editor's response was skeptical, though indirect. Without referring to Pitt, he wrote that he feared many years would pass before anyone designed a practical reaping machine.

The world's first patent on a mechanical reaper was issued in 1799, in England, to Joseph Boyce of Mary-le-bone. Several more inventors secured reaper patents soon after Boyce's was granted. Most of the proposed reapers were push-type machines that attempted to duplicate the manual cutting motion of a scythe. A typical design from the early 1800s had a drum mounted on the front, with several scythelike blades attached to the perimeter of the drum. As the car moved forward, the drum rotated and the attached blades whirled around like revolving scimitars. Machines of this type eventually evolved into the modern-day Lawnboy. If simple mowing had been their purpose, these machines might have been more successful. A practical reaper, though, had to do two things: It had to cut the stalks uniformly *and* leave the grain in orderly piles for gathering. Early British rotary reapers, like today's lawn mowers, tended to scatter clippings every which way. Whatever labor they saved in reaping was more than outweighed by their making the binders' job nearly impossible.

One exceptional English design of the period was that of Robert

Meares of Frome, Somersetshire. In 1800 Meares patented a set of large garden shears that rode on wheels. The invention didn't work very well, but it looked impressive. Seven years later, a surveyor for the duke of Bedford, a man named Robert Salmon, designed a reaper that looked alarmingly like a portable guillotine. It was one of the first grain cutters to incorporate a reciprocating cutter bar. Other than that, Salmon's invention had little to recommend it.

More than thirty other reaping machines appeared in England or Scotland in the forty-five years between 1786 and 1831, when Cyrus McCormick's machine made its debut in America. Among the most interesting examples from this crop of English inventions was that of James Dobbs, an actor from Birmingham. The remarkable thing about Dobbs's reaper was not the machine itself but its inventor's way of introducing it to the public. Dobbs took out the following advertisement in the Birmingham *Gazette* of October 10, 1814:

> J. Dobbs most respectfully informs his Friends and the Public, that having invented a Machine to expedite the Reaping of Corn [small grains], etc., but having been unable to obtain the Patent till too late to give it a general inspection in the field with safety, he is induced to take advantage of his Theatrical Profession and make it known to his Friends, who have been anxious to see it, through that medium. . . . To conclude [the performance] will be presented the celebrated farce of Fortune's Frolic. The part of Robin Roughhead will be taken by Mr. Dobbs, in which he will work the Machine in character, in an Artificial Field of Wheat, planted as near as possible in the manner it grows.

A review of Dobbs's performance appeared one week later in the same newspaper:

> Mr. Dobbs's newly invented Patent Reaping Machine was shown at our Theatre on Friday evening, and appeared to be highly approved of. Mr. D's first experiment was completely successful and we have no doubt the other could have been equally so, had not the scenery obstructed the progress of the machine, which, causing a little embarrassment, prevented Mr. Dobbs from working it so effectively as he could have

> wished. Mr. D's explanation of the principles and properties of this invention was very satisfactory, and, we are inclined to think, it will prove of great public utility.

IF CYRUS MCCORMICK had a serious transatlantic rival in his claim to have invented the first practical reaping machine, that rival was not James Dobbs but Patrick Bell. In 1826, as a twenty-five-year-old divinity student at the University of St. Andrews in Edinburgh, Bell experienced an epiphany of an unexpected kind. On an evening walk, he saw a pair of gardener's shears protruding from a hedge where their owner had apparently left them for the night. Suddenly a vision appeared to Bell: In his mind's eye, he imagined a reaper made from a row of similar shears that were operated mechanically.

Theology and reapers vied for Bell's attention for the next few years. He kept his work on the reaper a secret until its first public trial in 1828. First he built a few scale models, then his brother helped him build a full-scale, working prototype. When it was time to test the machine, the Bell brothers surreptitiously carted loose dirt into the shed where they were working on the reaper, covering the floor six inches deep. They then stuck a sheaf of oats in the dirt, one stalk at a time. The machine cut the artificial crop well enough but scattered the stalks in a hopeless mess. Anticipating McCormick by four years, Bell then added a reel to his reaper. He also added a canvas conveyor belt that collected the cut stalks and deposited them neatly on one side.

By the summer of 1831, when Cyrus McCormick's reaper made its first public showing at Steele's Tavern, at least seven Bell reapers were operating in Scotland. They harvested a total of 219 acres that year, averaging around 10 acres each per day. Ten acres a day was as much ground as three or four men with cradles could reap. A number of Bell's machines turned up in Europe, Australia, and America, becoming the first reapers ever exported for world trade. But Bell had no desire to profit commercially from his invention. He finished his studies at St. Andrews and spent the rest of his life as a minister and teacher. He never patented his reaper, saying that he "wished the implement to go into the agricultural world free of any extra expense." Without a single-minded promoter to champion it, Bell's reaper soon faded into oblivion.

Later generations of schoolchildren might have learned that

Patrick Bell invented the reaper if conditions in Britain had been more receptive to such a machine. But it found little local demand, mainly because harvest labor in Britain was cheap. A population explosion in Ireland had generated a seasonal tide of migrant Irishmen who would work for little more than a stable to sleep in and a few scraps of food. Labor, then, was readily available, whereas land was relatively scarce. The agricultural land in Great Britain tended to be hilly and carved up into small, irregularly shaped fields. An expensive machine that dispensed with inexpensive labor and worked best on large, flat fields made little economic or practical sense.

The situation was different in America. There it was land that was abundant, and labor that was scarce. Everything was growing: America's population, its cities, its economy, and the amount of territory within its borders. Barriers that in Europe stood in the way of extending the Industrial Revolution to agriculture did not exist in America. The economic equation was one such barrier; another was the legacy of feudalism: Even if new and better ways of farming could be found, European peasants lacked both the means and the incentive to adopt them. In America, however, yeomen farmers owned their own land. Much of that land, especially in the newly opened territories acquired in the Louisiana Purchase (an area more than twice the size of France and Spain combined), was excellent for farming. And the demand for surplus food and fiber, beyond what farmers could produce for themselves, swelled as eastern cities grew and western towns multiplied. Cities and their hinterlands developed in tandem, like Siamese twins—vitally linked partners in an expanding market economy. In America, conditions were ripe for the reaper.

THE FIRST PERSON to make, patent, and sell a reaping machine in the United States was not Cyrus McCormick but Obed Hussey. If Hussey hadn't existed in the flesh, Herman Melville might have had to invent him. Born in Maine, Hussey soon moved with his family to Nantucket, where the young Quaker decided, like Melville's Ishmael, to "sail about a little and see the watery part of the world." Also like Ishmael, Hussey's sea travels took him to the Pacific, where he rowed whaling skiffs in pursuit of the great cetaceans. It may have been during a whaling trip that Hussey lost an arm and the sight in his left eye. His photograph presents a man of arresting though forbidding appearance. A

Obed Hussey, Cyrus McCormick's archrival in the reaper wars.
(American Society of Agricultural Engineers)

square black patch covers his left eye, while his right eye stares out with a cold intensity, like the eye of a bird. The Mona Lisa–like expression on his craggy, beardless face can be read as conveying either an earthy sense of humor or a wicked arrogance. The historical record leaves his personality open to widely varying interpretations. To his friends he was the picture of Quaker beneficence. But to his enemies—and above all to Cyrus McCormick—he was a fiercely tenacious, one-eyed devil.

The little that is known of Hussey's early life reveals that on leaving the sea, he traveled inland to Cincinnati and took up inventing. He made a machine for stamping out pins and fish hooks, and another that molded candles mechanically. In 1832, when he was forty years old, a friend is said to have asked him, "Hussey, why don't you invent a machine to reap grain?"

"Are there no such machines?" Hussey replied.

"No," the friend assured him, "and whoever can invent one will make a fortune."

Hussey knew little about farming. Nonetheless, he set his mind to the principles of mechanical reaping and designed a workable machine in short order. He patented his reaper late in 1833. The following spring, *Mechanics' Magazine* printed a woodcut of Hussey's reaper, along with a description.

Cyrus McCormick saw the notice and went ballistic. He had not yet taken out a patent on his own reaper and worried that Hussey might gain rights to ideas that he, McCormick, had thought of first. He promptly fired a cannonball across Hussey's bow in the form of a seething letter to the editor.

> Rockbridge, Virginia, May 20, 1834
> Dear Sir:
>
> Having seen in the April number of your "Mechanics' Magazine" a cut and description of a reaping machine, said to have been invented by Mr. Obed Hussey, of Ohio, last summer, I would ask the favor of you to inform Mr. Hussey, and the public, through your columns, that that principle, viz., cutting grain by means of a toothed instrument, receiving a rotatory motion from a crank, with the iron teeth projecting

Hussey's reaper, 1833. (Smithsonian Institution)

> before the edge of the cutter for the purpose of preventing the grain from partaking of its motion, is a part of the principle of my machine, and was invented by me, and operated on wheat and oats in July, 1831. This can be attested to the entire satisfaction of the public and Mr. Hussey, as it was witnessed by many persons; consequently, I would warn all persons against the use of the aforesaid principle, as I regard and treat the use of it, in any way, as an infringement of my right.

Hussey chose not to reply. He was already selling reapers, although not in Virginia. His turf was farther north, in Maryland, New York, and Ohio. McCormick's distant saber rattling didn't faze the old sea dog. McCormick, for his part, rushed frantically to patent his machine, though he was soon sidetracked on an unrelated business venture with his father. The business, a smelting furnace, failed after iron prices collapsed in the panic of 1837. McCormick was left with little but debts and his reaper patent. The life of a farmer held little appeal for him, so he turned his attention back to the reaper, which he saw as his only prospect. He made his first two sales in 1839, in time for the harvest of 1840. Both machines failed miserably. About that time, Hussey boasted that "I consider myself alone successful. . . . Every previous attempt has totally failed and gone into oblivion."

This time it was McCormick's turn to ignore the bluster. He turned his attention to making some badly needed improvements on his machine. He made seven improved machines in time for the harvest of 1842; six of them sold. They cost him about $50 each to make, and he sold them for $100 (the equivalent of about $1,500 in today's money). This was as much cash as passed through the hands of many a pioneer farmer in a decade. In 1842, only a well-to-do planter with substantial grain acreage and a penchant for new, unproven devices would consider buying a $100 reaping machine. Fortunately for McCormick, that year's batch of reapers performed as advertised. William Peyton, a prominent Roanoke planter who bought one of the six machines, was so pleased with his purchase that he wrote a letter of commendation to the Richmond-based *Southern Planter.* Peyton said he thought McCormick's invention saved at least a bushel of grain per acre that would normally have fallen to the ground when cradle scythes were used. (One of Hussey's customers expressed a similar view when he remarked that the reaper "cheated the hogs.") And since it eliminated the need for eight laborers, Peyton said, McCormick's reaper could probably pay for itself in a single year if it was used on a sufficiently large acreage.

While McCormick was enjoying his first taste of success in Virginia, Hussey had established a reputation to the north and began expanding his territory southward. He wrote a provocative letter to the *Southern Planter,* referring archly to "another reaper in your state which is attracting some attention." He challenged this "other machine" to a head-to-head bout. Sensing the material for some lively copy, the editor of the *Southern Planter,* Charles Botts, arranged a contest. He persuaded a farmer named Ambrose Hutchinson, whose estate lay near Richmond, "to allow Hussey and McCormick to race their machines in his field for the entertainment of the country-side and for the best interests of agriculture."

The contest ended indecisively. McCormick's machine cut seventeen acres of grain compared with Hussey's two, but bad weather and a washed-out bridge had delayed Hussey's arrival, putting him at a considerable disadvantage. A rematch was set for a week later. This time, hoping to prove the superiority of his machine in tangled and downed wheat, Hussey attacked a particularly nasty patch of grain at a fast trot and broke his cutter bar. Once again, the judges felt that Hussey had been the victim of bad luck and declined to declare a winner. The enduring significance of the Richmond campaign lies not in who won or lost but in the spectacle of two rival inventors meeting in a field of battle to match their machines head-to-head. This was industrial competition at its most primitive and direct.

McCormick and Hussey's square-off in 1843 was the opening battle of the reaper wars. The two generals had soon enlisted armies of sales agents, who, in turn, led campaigns across an expanding theater of engagement—first in the wheat-growing regions of Virginia and New York, then in the Ohio River valley, followed by the prairie states of the Great West. As the reaper caught on among farmers, a growing number of manufacturers joined the fray until, shortly after the Civil War, more than 200 makers fought one another, sometimes violently, for a share of the market. The reaper wars escalated in ferocity, and companies began to fall like soldiers at Gettysburg.

The field trial was to the reaper wars what the Napoleonic charge was to the Civil War: the basic mode of combat. Reaper trials were often arranged by sales agents from a single company, who would rig the contest to show their own machine to best advantage while exposing the competition to maximum humiliation. A common tactic was for a manufacturer to pit one of its own brand-new, carefully tuned reapers against an old, broken-down machine made by a rival. Sometimes the rival's agents got wind of such a rigged trial and showed up to defend their company's honor. Knives and guns occasionally appeared to help decide the outcome of these confrontations.

Reaper trials often assumed a gladiatorial aspect. The ultimate test was to chain two reapers of different makes together, back-to-back, then hitch up teams of horses to pull them apart. The reaper that suffered the most damage lost. One popular variation on this theme was the horse-drawn demolition derby. Sales agents representing various reaper manufacturers were assigned to drive an opponent's machine in a torture test. The contestants deliberately ran the machines into fence posts or thick stands of weeds, causing spectacular crack-ups. Even in more creditable trials, acts of sabotage were frequent.

No reaper sale was ever final, not if the competition could help it. Some salesmen specialized in reversing rivals' sales after they were made. Such an agent would call on a farmer who had recently purchased another company's reaper (sales were nearly always on credit) and use every trick he knew to convince the farmer that he had bought a lemon. There was a story of one farmer who bought a reaper and woke up the next morning to find that his new machine had disappeared. A reaper made by another company stood in its place. The most unscrupulous reaper firms maintained "wrecking crews" that roved the countryside, tampering with enemy machines by night to undermine customer satisfaction. "Competition grew so severe and unbusinesslike," Cyrus McCormick, the inventor's grandson, wrote in 1931, "that the members of the leading organizations became enemies, not in a personal sense, but, like soldiers, believing that the thing had to be done. Anything the old-time men could do to the other fellow to knock him out was considered legitimate."

William N. Whitely, better known as "Wild Bill," was the Ben Hur of reaper warriors. He invented a combined reaper-and-mower he called the Champion. Whitely loved field trials, and pitted Champions against rivals at every opportunity. Nobody messed with Wild Bill Whitely. He had the physique of a prairie Paul Bunyan and a charismatic personality to match. "I've seen Bill Whitely racin' his horses through the grain and leanin' over with his long arms to pick the mice's nests from just in front of the knife," one Ohio farmer recalled.

The event that launched Whitely into the realm of legend occurred during a field trial in Jamestown, Ohio. Whitely was running neck-and-neck with a competitor, Benjamin Warder, when Wild Bill sprang off his Champion, unhitched one of his two horses, and jumped back on the reaper to resume cutting with a single horse. Seeing this, Warder did the same. Whitely then jumped off his machine again and unhitched the remaining horse. He put the horse's collar around his own neck and pulled the Champion a full length of the field, laying down a perfect swath of cut grain behind him. Warder conceded. More than 500 spectators witnessed the event, which was widely reported in the press. On the strength of this and other Whitely publicity stunts, the Champion became, for a while, the best-selling reaper in the country. Springfield, Ohio, where Champions were made, earned the nickname "Reaper City," as it turned out as many as 160,000 machines a year.

Reaper competition grew so fierce and chaotic that it undermined the stability of the industry. Nobody was making a profit. Several of the most prominent company heads called a truce and met to discuss their common predicament. A chairman was appointed, and he opened the meeting by asking, "What ought we to do to improve the conditions of our trade?" No one said anything for a moment, then a mild-natured man named John P. Adriance calmly said, "Kill Whitely."

CYRUS McCORMICK LACKED the popular appeal of a Wild Bill Whitely, but he excelled in tactics of his own making. He became one of the first manufacturers to advertise widely in agricultural magazines, writing all his own ad copy and elevating the practice to a craft in itself. He was the first to sell reapers on a time-payment plan, and so established himself as a pioneer of revolving credit. Mainly, though, he had a good nose for the business, anticipating demand and making changes to meet it.

The smartest thing he ever did in this respect was to move his business west. The Valley of Virginia clearly wasn't in the running as the reaper capital of the world. It was geographically isolated, its fields were stumpy, rough, and small, and its soil (like farm ground all over the East) was nearing exhaustion after decades of cultivation with no compensating fertilizer and little crop rotation. Eastern grain crops, weakened by nutrient-depleted soils, were susceptible to attack from diseases and insects. The Hessian fly decimated eastern wheat fields during the mid-1830s, contributing to a bread shortage that erupted into urban rioting. The Erie Canal (opened in 1825) linked new grain-growing regions in the West with eastern markets. The old grain-growing districts of Virginia and New York, with their depleted soils, could no longer compete. McCormick ventured to Wisconsin, Illinois, and Missouri in 1844 to see the new farming districts for himself. Writing home, he spoke of vast, open fields and of bumper crops wasting away because there was insufficient labor to harvest them. He saw clearly that the future of the mechanical harvester lay in the West. He began to think of himself as an ambassador of prosperity, an agent for America's westward march of progress. As a sometime farmer himself, he liked to think that his machine would help the American farmer "beat the world growing wheat."

McCormick moved his base of operations to Chicago in 1847. Chicago turned out to be a brilliant choice, though it was by no means the most obvious choice at the time. Cincinnati and Pittsburgh were bigger and had better manufacturing facilities. St. Louis, with its ideal location at the confluence of the Mississippi and Missouri rivers, was the western center of waterborne trade, and steamboats were still the main vehicles for grain transport at that time. Chicago, by contrast, was still a swampy, somewhat sleepy town of fewer than 17,000 inhabitants in 1847. McCormick could hardly have known it, but the town was on the verge of taking off. The year after McCormick's move, Chicago acquired its first railroad, grain elevator, telegraph line, canal, and stockyard. That same year, a group of merchants founded the Chicago Board of Trade. Whether guided by extraordinary luck or uncanny instinct, McCormick found himself perfectly situated. His business prospered immediately, and from the time of his move westward he became the reaper king. He called his machine the Virginia Reaper, in honor of its birthplace. The steam-powered factory that made them was considered one of the marvels of Chicago.

Obed Hussey's business instincts served the ex-seaman less well. He was a fierce competitor, but he had certain scruples. He refused, for example, to add a reel to his reaper even though he knew it would improve the machine's performance. The reel, he said, was McCormick's invention, and he wasn't about to steal another man's idea. (Mc-

Cormick, on the other hand, did not hesitate to adopt Hussey's superior design for a cutter bar, which is basically the design still in use today.) Even as Hussey was losing his battle against the shrewd marketer of the Virginia Reaper, the one-eyed Quaker declared, in 1848, that "I alone have been successful with a reaper in the United States and now believe myself without a rival in any country." That was his swan song in the reaper business. Within a few years he had sold his reaper patents and set out to design a machine that, he thought, would surely be the next great innovation in agriculture: the steam plow. Hussey's steam plow, however, never amounted to much. The inscrutable whale hunter, who was variously described as "poetic," "whimsical as the weather," "a picturesque debater," and a man of "extremely sensitive, modest and unassuming personality," was killed when he fell between the moving cars of a train in 1860. He had left the train to fetch a glass of water for a thirsty child and lost his balance when the cars lurched forward as he was reboarding.

Cyrus McCormick was fashioned from a different mold. He was an intense, severely sober man, so focused on his business that little else

Cyrus McCormick, as photographed by Mathew Brady in 1860. (International Harvester Collection, Wisconsin Historical Society)

could intrude on his attention. He avoided marriage until he was almost fifty years old. Physically, his was a commanding presence. He stood about six feet tall and had alert, dark eyes, dark hair that grew in untamable swirls, and a full, wavy beard. His girth swelled with his fame and fortune. He was a fastidious dresser and, like many farm boys who chose white-collar occupations, he was proud of his clean, well-manicured hands. He kept to himself most of the time, preferring to read his Bible (he was a devout old-school Presbyterian) rather than engage in small talk. He indulged in neither alcohol nor tobacco. Once, when a friend kidded him about his clumsy handling of some personal matter, McCormick whirled in his chair and thundered, "I have one purpose in life, and only one—the success and widespread use of my machines. All other matters are to me too insignificant to be considered."

If McCormick ever knew a moment of self-doubt, he didn't show it. The motto on his family's ancient coat of arms, *"Sine Timore"*—"Without Fear"—was engraved deep in his bones. "His brain had certain subjects distinctly mapped out," wrote Herbert Casson, one of McCormick's admiring biographers. "What he knew—he knew. He had no hazy imaginings. He lived in a black and white world and abhorred all half-tints. He was right—always right, and the men who opposed him were Philistines and false prophets, who deserved to be consumed by sudden fire from Heaven." An attorney who once worked for McCormick recalled that "the exhibition of his powerful will was at times actually terrible. If any other man on this earth ever had such a will, certainly I have not heard of it."

McCormick was extremely litigious. When, in 1848, the U.S. Patent Office turned down his request to renew his early reaper patents, he fought the decision with Pyrrhic persistence. The obsolete patents were of little value to his business, but of great importance to his pride: He felt they established his claim as the father of the mechanical reaper. The case went all the way to the floor of the U.S. Congress, where a vote went against McCormick.

He could be just as stubborn over smaller matters. Once during a railroad trip, a clerk charged him a fee for excess baggage. The deeply Scottish industrialist refused to pay, insisting that the fee was too high. The clerk stood his ground, so McCormick stormed off the train at the next stop, sans baggage, and sued the railroad. (His trunks wound up back in Chicago, where they went unclaimed and were eventually destroyed.) Twenty years later, shortly before his death, McCormick won on final appeal. The case had cost him five times as much as the damages his estate collected from the railroad, but that hardly mattered. What counted more to the reaper king was that his Calvinist sense of righteousness had been vindicated.

McCormick considered the reaper to be his proprietary invention, so he sued newcomers to the industry almost as a matter of routine. One such battle, *Cyrus H. McCormick v. John H. Manny,* became one of the most celebrated court cases of the age. In the summer of 1852, McCormick took great offense when, at a New York agricultural fair, Manny won first prize for his horse-drawn mower and second prize for his reaper, while McCormick's machines failed to place. Worse, Manny had had the effrontery to build a factory in Rockford, Illinois, challenging McCormick on his own turf. McCormick filed a Bill of Complaint in U.S. Circuit Court charging that Manny had infringed two minor patents. The stakes were high: If McCormick won, every reaper maker in the country would be forced to pay tribute to the reaper king in the form of patent royalties. By this time, scores of reaper manufacturers, large and small, had established themselves across the north, mainly in the Midwest. Many of these manufacturers, which had previously fought one another like Hatfields and McCoys, found common cause against McCormick and rallied around Manny. Both sides hired all-star legal teams. McCormick retained Edward N. Dickerson, a distinguished patent attorney from New York, Reverdy Johnson, a leader of the American bar from Maryland, and William H. Seward, another expensive lawyer from New York. Manny retained George Harding, a prominent Philadelphia patent attorney, and Harding's partner, Edwin M. Stanton. Both the press and the public looked forward to a diverting courtroom battle.

Since Manny's lawyers thought they would be arguing before an Illinois judge, his attorneys figured they should find a member of the Illinois bar to act as their front man. Someone suggested an obscure circuit lawyer from Springfield named Abraham Lincoln. Lincoln was said to know the presiding judge, so the defense team hired him, sight unseen. On meeting their junior colleague, however, Harding and Stanton were dismayed by the man's unpolished speech and awkward appearance. They were relieved when the venue was changed to Cincinnati, where Stanton could argue the case himself. Stanton performed brilliantly, and the judge ruled for the defendant, John Manny.

The case marked a turning point in Lincoln's career. Thereafter, he was determined to prove himself the equal of college-educated, big-city lawyers like Stanton. Lincoln went home to Springfield with a generous $1,000 fee in his pocket (Stanton received a princely $10,000). A portion of the money helped build the new house that Mary Todd Lincoln had her heart set on, now the Lincoln Home National Historic Site in Springfield. The remainder of the fee later supported Lincoln during his debates with Stephen Douglas, thereby opening Lincoln's path to the White House.

DESPITE A FEW legal disappointments, by mid-century Cyrus McCormick was well on his way to making his first million dollars—the dream that had seemed so implausible fifteen years earlier. By 1850 he was selling about 1,500 machines a year to farmers from New Jersey to California. Commercial agriculture was supplanting Jeffersonian subsistence in areas where new transportation corridors created access to growing markets. American farmers grew half again as much wheat and corn in 1850 as they had in 1840. With the completion of the first midwestern railroad lines and a canal between the Illinois River and Lake Michigan (connecting the vast Mississippi watershed with the Great Lakes), Chicago was rapidly emerging as the transportation hub of the Great West. As Chicago's fortunes rose, so did Cyrus McCormick's. He was the largest reaper manufacturer in the United States. But competition was heating up. Instinctively, he looked around for new realms to conquer—new markets where he could widen his lead over all challengers and assure his claim to the reaper crown for years to come. He fixed his gaze across the Atlantic, where Great Britain and Europe glimmered as unclaimed prizes.

IN JUNE OF 1849, the *Chicago Daily Journal* ran a story under the headline "GEM OF THE PRAIRIE." It described a gorgeous reaper that had been specially built at Cyrus McCormick's factory. The reaper's woodwork was made of fine Michigan ash, its iron gears and fittings were plated with polished brass, and its canvas-covered grain platform was emblazoned with a painted American eagle, "talons, bolts, and all." McCormick planned to present this reaper to Prince Albert, patron and president of the Royal Agricultural Society. He hoped that this ceremonial introduction would open the door to a British reaper market.

But an even better opportunity to introduce the reaper to Britain soon presented itself. Prince Albert declared his intention to host an international exhibition of industrial arts—the first world's fair. A great building was needed to accommodate the exhibition. Self-taught gardener and architect Joseph Paxton won the commission with a giant conservatory of iron and glass. *Punch* dubbed Paxton's design "The Crystal Palace," and the name stuck.

Of the nineteen acres the Crystal Palace enclosed, British exhibits would take up half; the rest of the world was invited to fill the other half. Here was a chance for the United States to astonish Europe with the quantity and quality of its industrial arts. American representatives reserved 85,000 square feet, or nearly two acres. No one knew for sure how all that space would be filled, but the U.S. Congress encouraged exhibitors by promising free shipping for their displays. McCormick volunteered to exhibit his gilded reaper.

The Great Exhibition began on May 1, 1851. Queen Victoria took part in an opening ceremony that was said to surpass her coronation in splendor. Those who entered the Crystal Palace confronted a spectacle that was almost too wonderful to believe. A fountain made of cut glass, twenty-seven feet high, gushed a crystalline shower near the main entrance. Beneath the fanlight glazing of the central arch, a row of large elms (spared during the Palace's construction) spread their boughs. More than eight miles of display tables covered the ground floor and galleries. Exhibits ranged from minor curiosities to objects of astonishing craftsmanship and beauty. Some represented fundamental new discoveries in applied science.

The British proudly displayed Montgolfier's hydraulic ram, a 700-horsepower marine steam engine made by James Watt & Co., a locomotive capable of going sixty miles an hour, and a detailed scale model of a proposed canal across the isthmus at Suez. The host country exhibited exquisite pottery from Wedgwood and Doulton, ceremonial cutlery from Wilkinson, sumptuously packaged preserved fruits from Fortnum & Mason. Velvet pile tapestries, plush carpets, fine linen and sailcloth hung from the building's slender iron girders. The incomparable jewelry and gem collections of Henry Hope and his son, A.J., were eclipsed only by the breathtaking sight of the Koh-i-Nor diamond, still in the rough.

Austria sent a suite of five rooms furnished and decorated by Leistler of Vienna, the grandest and most expensive interior decorator of the day. From Prussia came Siemens's early electric telegraphs; from

Munich, colossal bronze lions by Ferdinand Müller. France sent the most magnificent tapestries, carpets, and porcelains in the building. Russia exhibited sixteen-foot-high vases made of malachite, massive urns carved from jasper, and clocks made of malachite and ormolu. From the Czar's own closets came a gleaming cloak of silver fox valued at 3,500 pounds sterling.

The United States was unable to muster enough objects to fill its assigned space. British newspapers ridiculed the sparse and barren American exhibit, calling it the "prairie ground." Surrounded by the silks, statuary, and jewels of Europe, the United States displayed such items as false teeth, artificial legs, air-evacuated coffins, corn-husk mattresses, and chewing tobacco. From Ohio there was a "stuffed buck-eyed squirrel" and some Catawba wine. Among the more impressive American entries were items made of India rubber from the Goodyear Company, and firearms (including a revolving six-shooter) made by Samuel Colt. But the most popular object in the American display—in fact, one of the most avidly sought-out exhibits in the entire Crystal Palace—was the *Greek Slave,* Hiram Powers's life-size statue of a naked woman in chains.

McCormick's eagle-emblazoned reaper occupied a central place on the "prairie ground." Hussey had also sent one of his reapers. On the opening day of the Great Exhibition, the London *Times* described McCormick's entry as "a cross between a flying machine, a wheelbarrow and an Astley chariot." Other papers called it "an extravagant Yankee contrivance," or "huge, unwieldy, unsightly and incomprehensible." The *Times* article continued: "Other nations rely upon their proficiency in the arts, or in manufactures, or in machinery, for producing effect. Not so with America. She is proud of her agricultural implements which Garrett, or Ransome and May [famous British inventors], would reject as worthless; she is proud of her machinery, which would hardly fill one corner of our Exhibition."

The Great Exhibition lasted 140 days and was a huge success. In the middle of the show's summer run—on July 24, to be exact—McCormick's "extravagant contrivance" was missing from its spot in the Crystal Palace, as was Hussey's reaper. The Exhibition's Jury on Agriculture had taken them to a farm about forty-five miles from London, at Tiptree Heath, to pit them against each other and a third reaper of English make. Neither McCormick nor Hussey was present; both planned to show up in England later in the harvest season.

Horace Greeley was on hand for the trial, though, along with 200 other spectators. Greeley described the day as "sour, dark, drenching." The wheat was not only wet, it was also on the unripe side. McCormick & Company had sent an American mechanic to England to baby-sit its reaper during the exhibition, and this fellow, whose name was McKenzie, was assigned to operate it during the trial. Hussey's machine was attended by an Englishman who had never seen a reaper work before. He didn't know how to adjust the machine's platform and found it impossible to rake off the grain. He didn't know what to do when the Hussey cutter bar clogged. Things went even worse for the British machine—it couldn't even get started in the wet, unripe grain. McKenzie, meanwhile, drove the Virginia Reaper smartly through the field. The judges estimated that it cut at a rate of twenty acres a day. A cheer went up for McCormick.

A few days later, a McCormick sales agent reported home from the floor of the Crystal Palace. "The 'Prairie Ground' is now thronged," he wrote. "McCormick's machine is put back in its place and I believe yesterday more visited it than the Kohinoor diamond itself." Greeley congratulated McKenzie for taking "the wind out of John Bull's sails."

But the Jury on Agriculture wasn't satisfied. They wanted to test the reapers again in fair weather, so a second trial was set for August 6. While Obed Hussey was dithering in France, Cyrus McCormick showed up in England just in time for the contest. After a "thorough trial" on the farm of Philip Pusey, M.P. (whom Benjamin Disraeli described as "one of the most distinguished country gentlemen who ever sat in the House of Commons"), the Virginia Reaper triumphed again. The jury gave it a Council Medal, the highest honor bestowed at the exhibition.

The tone of the press suddenly changed. "The reaping machine from the United States is the most valuable contribution from abroad, to the stock of our previous knowledge, that we have yet discovered," wrote *The Times,* somehow forgetting its previous contempt. In October, after the Crystal Palace show was over, *The Times* judged in retrospect that the McCormick reaper alone could "amply remunerate England for all her outlay connected with the Great Exhibition." Editorials now exhorted British farmers to adopt reapers and other horse-powered machines to ensure their own economic survival, and to reduce Britain's dependence on other nations for food.

. . .

By mid-century, Britain had become the nucleus of an expanding global grain trade. Revolutionary advances in steam power, ship design, canal building, and railroads changed the economics of shipping bulky commodities long distances, which had the effect of connecting markets that had previously been isolated. As the world seemed to get smaller, events in one part of the globe tended more and more to trigger secondary and tertiary effects in places thousands of miles away. In 1853, for example, war between the Russian Empire and the Ottoman Empire (which received support from British and French forces) broke out on the Crimean Peninsula. The hostilities, which lasted for three years, interrupted the flow of grain from Russia to Western Europe. As a result, American wheat prices immediately shot up by 50 percent. In Chicago, the up-and-coming transportation hub of the Great West, grain shipments tripled during the Crimean War.

The guns firing in Sevastopol and Balaklava had a salutary effect on production at the McCormick factory, 6,000 miles away. As the demand for grain increased, so did reaper sales. The American grain belt expanded farther and farther to the north and west. A recession in the U.S. economy in the late 1850s barely put a dent in reaper sales as western settlement continued and farms proliferated. Some western boosters claimed that "rain followed the plow," a ridiculous theory that was soon disproved. It *was* true, however, that railroads followed the reaper. As new wheat districts were settled, railroads were built to haul the grain to market, and the proceeds helped pay for the railroads. The reaper (along with improved plows and planters) made this agricultural expansion possible, even as labor remained scarce. William Seward, McCormick's sometime lawyer and Lincoln's secretary of state, once remarked that the reaper "pushed the American frontier westward at the rate of 30 miles a year."

The king of American agriculture at mid-century was not wheat, however, but cotton. The United States produced three-quarters of the world's supply, much of it exported to England's "dark Satanic mills" (as William Blake called his country's sulfurous, coal-blackened textile factories). Cotton was America's single largest export, accounting for more than half of all income from foreign trade. Cotton, however, was a southern crop, the product of long, hot summers and slave labor. After the Union dissolved, the North needed a "cotton" of its own to generate foreign exchange and sustain its growth.

Clearly, the cotton of the North was grain. The antislavery Democrat Edwin M. Stanton made the comparison explicit in 1861:

> The reaper is to the North what slavery is to the South. By taking the place of regiments of young men in the western harvest fields, it releases them to do battle for the Union at the front and at the same time keeps up the supply of bread for the Nation and the Nation's armies. Thus, without McCormick's invention, I fear the North could not win and the Union would be dismembered.

The Civil War stepped up pressure on reaper manufacturers to make the "slave of the North" ever more efficient. Originally, both McCormick's and Hussey's reapers needed two people to operate them: one to drive the horses, another to walk alongside and rake piles of grain onto the ground. The first major improvement (which McCormick and other reaper makers introduced around 1850) was simply the addition of a seat to allow the raker to ride rather than having to walk along beside the machine. Adding a raker's seat now seems like a simple and obvious refinement, yet at the time it posed several technical challenges. The weight of the seat and its rider unbalanced an already asymmetrical machine and tended to exacerbate the problem of side draft (a pronounced pulling to the right, the side from which the cutter bar protruded). Once the problems of balance and support were solved, the raker's seat speeded up the entire operation, since the pace of the machine was no longer limited to the speed of a walking human.

The next big step was to eliminate the raker altogether. Self-raking reapers of various brands and designs began to come out just before the Civil War, and the acute shortage of farm labor caused by the war helped to speed the adoption of this refinement. Most self-raking reapers substituted a system of rotating, windmill-like rakes for the reel. If the early reapers had looked rather awkward and fragile, the first self-rakers looked absurdly so. Cyrus McCormick took his usual stance toward an idea that he had not pioneered: He scorned self-raking reapers as if they were the work of the devil. By 1860, though, his company's share of the reaper market had slipped to less than 10 percent. McCormick was forced to swallow his scorn and license outside patents so that his company could market a self-raking reaper of its own.

The self-raking feature cut the number of people needed to operate a reaper in half, from two to one. Still, a crew of four or five human binders was needed. The binders walked through the field, gathering the bundles of grain the reaper left in its wake and tying bands of straw firmly around them to form sheaves. The sheaves then had to be stacked end-on-end in shocks to dry. (Grain was reaped while it was still slightly green so the kernels wouldn't pop out onto the ground when the cutter bar shook them.) After the grain had dried, the shocks were disassembled and the sheaves taken to another location for threshing. It was all very well for one person to be able to reap twenty acres in a day, but the rest of the harvest procedure was still very labor-intensive, demanding dozens of hands at every step.

But what if a reaper could be made to perform the function of the binders as well? The idea seemed utterly fantastic.

Binding was one of the more tedious jobs not just of the harvest but in the whole seasonal cycle of farm work. Charles Marsh, an Illinois farmer, hated it more than anything. Binding required little skill but much stooping. His back ached at the thought of it. As he reaped his grain in the summer of 1857, an idea took shape in his mind. He imagined taking an ordinary reaper and adding a moving canvas belt that would carry the cut grain up an incline and deposit it on an elevated table. One or two band tyers could ride on the machine and tie sheaves while they sat. Before long, Charles and his brother had translated this idea into a working machine. They discovered that with practice, two band tyers riding on the reaper could outpace four on the ground. They called their machine the Marsh Harvester.

Other people called it a man-killer. Binding the old-fashioned way was tedious, but it could be done at a leisurely pace. To keep up with the flow of grain on a Marsh Harvester, however, the band tyers had to work like Rumpelstiltskin. Charles Marsh tried to dispel the idea that his invention was impractical. He arranged public demonstrations to show how easy and pleasant harvesting the Marsh way could be. To shame his male detractors, he recruited two women to sit on the binding platform. During demonstrations the women smiled, waved, and left a trail of bound sheaves in their wake without breaking a sweat (they were, it turned out, experienced grain binders from Bavaria). At other times, Marsh asked for volunteers from the audience to try their skill on the binding platform. Invariably he chose the most forlorn-looking men in the crowd, hoboes who apparently hadn't done an honest day's work in years. These men would step up onto the platform and amaze spectators by tossing off sheaves like seasoned pros (which, of course, they were). But the real tour de force occurred during a demonstration in 1864, when Charles Marsh himself mounted the harvester and, with no assistance, cut and bound an entire acre in fifty-five minutes. This was a herculean feat that would have created a bigger sensation if Grant and Lee hadn't been down in Virginia, grabbing all the headlines.

Machines modeled after the Marsh Harvester captured an expanding share of the market for reapers during the decade following the war. A dominant figure in the manufacture of such machines was William Deering. Deering was a dry-goods merchant from South Paris, Maine, who journeyed to Illinois in the late 1860s. He had $40,000 that he wanted to invest in farmland. While in Chicago, Deering called on an old friend and fellow Methodist named Elijah H. Gammon, who had recently acquired the manufacturing rights to the Marsh Harvester but needed money to set up a factory. Gammon persuaded Deering to invest his $40,000 in a harvester business. Deering gave the money to Gammon and returned to Maine. Two years later, when Deering came back to see how his investment had fared, Gammon informed him that his $40,000 had tripled. Not long after that, Gammon fell ill and asked Deering to come to Illinois to manage the business.

"So," Deering recalled much later, "in that way I got into the harvester business and had to stay in. But I did not even know, at that time, the appearance of our own machine." Mechanical details weren't Deering's strength; instead, he brought a keen, dispassionate approach to the business. The chaotic reaper industry desperately needed an infusion of levelheaded business acumen. Deering was like an MBA in a Turkish bazaar. By 1875 he had bought out Gammon and built a factory in Chicago that turned out 6,000 harvesters a year, exceeding even McCormick's output. The starched, soft-spoken cloth merchant from Maine became the new reaper king.

McCormick & Company faltered during the 1860s and '70s, in part because its founder had discovered the charms of Europe. Feeling out of place as a closet Confederate in a Yankee city, McCormick lived as an expatriate during much of the Civil War. He particularly enjoyed socializing with European nobility and managed to get himself invited to the estate of Napoleon III at Châlons, where he chatted with the emperor while his Virginia Reaper was demonstrated on the wheat of Champagne.

Meanwhile, back in Chicago, Cyrus's brother Leander was running the business and feeling underappreciated. Leander eventually left McCormick & Company and published a sensational pamphlet bitterly asserting that Cyrus had stolen the ideas for his reaper from their father, Robert McCormick. Leander claimed, furthermore, that the person who had contributed most to the advancement of the reaper was Cyrus's old enemy, Obed Hussey.

One result of McCormick's social distractions and long-distance management style was that his company lost its edge as an innovator. McCormick was late in introducing a self-raking reaper, then failed to add a Marsh-type harvester to its line until a new development was about to make even the Marsh machine obsolete.

THE IDEA THAT a machine could imitate the intricate motions of a human hand in tying a knot seemed fabulous almost beyond belief during the first half of the nineteenth century. Then, on July 22, 1850, an Ohio inventor named John Heath became the first person to patent a machine that automatically bound sheaves with twine. It compressed the bundle of stalks, wrapped twine around the bundle, cut the twine to length, and presented the loose ends for tying. A human then had to intervene to make the knot. The device was simple, direct, and promising. One problem, however, was finding a suitable kind of twine to use in the binding machine. Should it be made of jute fibers? Manila? Sisal? Flax? The traditional material for hand-binding was straw, but straw proved too brittle and variable to work in an automatic knotter.

Six years later, one C. A. McPhitridge from Missouri introduced an inelegant way of mechanically consummating the tie that binds. He got around the complexity of knotting limp cord by using wire. Because of its stiffness, wire could be made to fasten around a sheaf with a simple twist. A number of wire-binding devices appeared during and immediately after the Civil War. McCormick and his underlings dismissed wire binders as "a swindle" and "sheer humbug." And they weren't alone in pooh-poohing the idea. Like twine, it was generally thought, wire was too scarce and expensive to be used in such profligate quantities. Farmers weren't yet used to the idea of buying commercial "inputs" to substitute for useful materials, like straw, that they could produce themselves.

Then, during the 1870s, barbed-wire fencing started carving up the treeless prairies and plains of the West, begetting a ready supply of inexpensive wire. Wire binders became a more practical possibility, and a number of inventors produced workable models. Among them were Sylvanus Locke, Charles Withington, James Gordon and his brother, John Gordon.

When these men applied for patents on their respective wire binders, the Patent Office realized it had a horrible mess on its hands. The patentees' claims overlapped, intertwined, and tangled like a snarled mass of barbed wire. Lawsuits for patent interference started zinging around like stray bullets. Unsure of how it would all turn out, some implement manufacturers signed contracts with more than one inventor. The McCormicks followed this course, deciding that they should get in on the divvying up of patent rights lest they find themselves out of the game when the chips fell. The McCormick brothers signed an agreement with the Gordon brothers, promising to pay the Gordons a royalty of $10 for every wire-binding machine the company made. Not long after that, however, the McCormicks decided they liked Charles Withington's wire-binding device better. They started manufacturing Withington wire binders in 1874, ignoring their contract with the Gordons.

The Gordons sued, joined by a partner named Osborne. Cyrus McCormick, approaching his seventieth birthday, had not lost his savor for a good legal battle. He fought tenaciously until he died, ten years later. His son, Cyrus H. McCormick, Jr., agreed to settle the Gordon-Osborne suit by paying the partners a back royalty of $6 for each wire-binding machine sold during the preceding decade. This came to $225,000—an enormous fortune at the time. Osborne and the two Gordons were summoned to collect their settlement. In *The Romance of the Reaper* (1908), Herbert Casson relates what happened next:

> When they called at the McCormick office in Chicago, they were taken to a small room on the top floor and shown a great pyramid of green currency.
>
> "There is your money," said McCormick's lawyer. "Kindly count it and see if it is not a quarter of a million dollars."

The three men gasped with mingled ecstasy and consternation. "B-b-but," stammered one of them, "how can we take it away? Can't you give us a cheque?"

"That is the right amount, in legal money, gentlemen," replied the lawyer. "All I will say is that there are a couple of old valises in the closet—and I wish you good afternoon."

For several hours Osborne and the Gordons literally waded in affluence, counting the money and packing it into the valises. By the time they finished, it was eight o'clock. The building was dark. The elevator was not running. They were hungry and terrified. Step by step they groped their trembling way downstairs, and staggered with their treasure through the perilous streets to the Grand Pacific Hotel. None of them ever forgot the terror of that night.

Cyrus McCormick gained a small measure of vengeance from the grave.

A twine binder, which automatically cut stalks and bound sheaves in a single pass. It seemed like magic. (National Archives)

Wire binders sold by the tens of thousands, but they were one of the most controversial farm implements ever invented. The reason was the wire. It clogged threshing machines, damaged milling equipment, and allegedly killed livestock when the animals ingested pieces of it. What if a piece of binding wire got into a loaf of bread? In theory, twine would be a far better binding material. But first there had to be a good knot-tying device. Then there had to be a supply of twine.

William Deering supplied both in a coup that startled the world in 1880. He had secretly manufactured 3,000 twine binders (using an ingenious knotter that John F. Appleby devised by combining the ideas of several other inventors) and ordered ten boxcar-loads of custom-made twine. Deering's twine binders all sold immediately, and the much-criticized wire binder suddenly became obsolete.

The story was told at Deering & Company that the first farmer to buy a twine binder took it to his field, hitched his horses to it, climbed onto the seat, and said, "What am I to do?" A Deering mechanic who happened to be on hand replied, "Do? Do nothing. Drive the horses!" The farmer obeyed, making a circuit of the field. He looked behind him and saw a row of beautifully bound sheaves. Shouting with amazement, the man swore there must be a genie hidden in the machine.

Binders (as the machines that both reaped grain and bound it into sheaves came to be called) proved to have great staying power. They remained the backbone of the grain harvest for more than a half century.

The binder seemed like a new and marvelous machine when my grandfather was born in 1888. He grew up around binders, and used a horse-drawn Deering model to harvest his crops until World War II. But by then the binder, too, had become obsolete, replaced by an even more marvelous machine called the combine.

One of the more memorable sales in the career of C. H. Haney, a salesman for the McCormick Harvesting Machine Company, took place in the summer of 1896 on the estate of Otto von Bismarck. The elderly unifier of Germany, as it turned out, was only weeks from death. He had heard about the American machines that not only cut wheat but bound sheaves as well. Bismarck was wary of the effect such machines might have on traditional Prussian farm life, of which he was a product. But curiosity got the better of him, and he asked for a demonstration.

"Bismarck sat in his carriage," Haney recalled a few years later, "but he ordered his driver to follow the harvester as closely as possible. He looked very old and feeble. For quite a while he watched me operating the machine. Then he made a sign for me to stop.

" 'Let me see the thing that ties the knot,' he said.

"I took off the knotter and brought it to his carriage. With a piece of string I showed him how the mechanism worked, and gave him a bound sheaf, so that he could see a knot that had been tied by the machine. The old man studied it for some time. Then he asked me, 'Can these machines be made in Germany?'

" 'No your excellency,' I said. 'They can be made only in America.'

" 'Well," said Bismarck, speaking very good English, 'you Yankees are ingenious fellows. This is a wonderful machine.' "

A GIGANTIC INVENTION

The peculiar ingenuity of the American has supplied the want of laborers, in a country where agriculture is carried on by wholesale, especially in the cereals, by an instrument of the most singular and elaborate construction.

—James Fenimore Cooper, *The Oak Openings* (1848)

KANSAS CITY'S ANSWER to the Crystal Palace is the Bartle Hall Convention Center, a drab concrete monolith that bestrides 12th Street, near City Hall, with all the civic grace of an interstate overpass. On a cloudy Sunday morning in January 1992, a tour bus stopped at Bartle Hall and dropped off thirty passengers, who made a beeline for the revolving doors. Among the passengers were Ralph Lagergren and Mark Underwood, who, for fun, tried to trip each other up in the revolving door by accelerating it to centrifuge speed. Inside the building Mark pulled at the knot of his necktie with evident discomfort. "I feel out of place in this suit," he said. "I only wear it for marriages and funerals."

"Which is this?" Ralph asked.

"I don't know yet," Mark replied. "Ask me tomorrow."

In fact, the event that had brought them to Kansas City was neither a wedding nor a funeral but a contest. Sponsored by the American Farm Bureau Federation, the nation's largest organization of farmers, the contest was like a science fair for farmer-inventors. Twenty applicants from around the country had won berths in a showcase of farm-bred ideas at the Farm Bureau's annual convention. That evening, at the end of the convention's first full day, a panel of judges would narrow the field of contestants to four finalists. The following day, at a general assembly where President George Bush was scheduled to make a speech, the grand-prize winner would be announced.

"I don't know if we'll win," Ralph said as he rode an escalator up to the exhibition floor. "But it would be great if we were chosen as finalists, at least. The finalists get to share a stage with the president of the United States. That's an opportunity you don't get every day. You can't beat that kind of exposure."

I asked what good publicity would do them at this stage, long before they had their machine ready for the market.

"We want to get the word out to as many farmers as we can so they'll start putting pressure on the big companies," Ralph answered. "If enough farmers show up at their local John Deere or International Harvester dealers and say, 'Hey, how about that Bi-Rotor combine?,' then that's what we want. It puts some heat on the big boys to give us a serious look. Also, in case somebody tries to get around our patents, we're in a stronger position if we've been featured in lots of articles and news stories. We want farmers to know that *somebody* is working on the combine of the future, and that somebody is *us*."

Mark was much less inclined to seek the limelight. He was wary of exposing his ideas to scrutiny, especially before they were perfected. He cared about nuts, bolts, and hydraulic drives, not publicity. But Ralph had convinced him to broaden his focus a little, to see a bigger picture. Mark now recognized that an event like the Farm Bureau contest could help their cause in the long run. Besides, it was fun to spend a couple nights at a nice hotel in Crown Center, maybe "do a sauna" or two, as he put it.

Far more than fun or publicity, though, Mark was after the grand prize of the Farm Bureau contest: a year's free use of a big blue Ford New Holland tractor. "It's got all their latest technologies on it," Mark said—adding, with a wicked grin, "I want to tear it down and figure out where the engineers went wrong."

The Farmer Idea Exchange candidates had set up their booths in one corner of a huge exhibition area in Bartle Hall. The contestants' corner was like a nursery for budding entrepreneurs. In a few years, with time and a lucky break, maybe the fledgling inventors would leap the blue velvet curtain and join the hundred or so commercial exhibitors whose displays and products filled the rest of the hall—the John Deeres, International Harvesters, and Ford New Hollands of the world.

The ideas on display in the Idea Exchange section ran the gamut from little gadgets to big machines, and the effect of all of them, crammed together in a villagelike cluster, was to affirm that the inventive spirit is thriving on American farms. There was Clarence Stonebrink of Enterprise, Oregon, showing a video of his self-propelled baling machine. A few booths up stood Tony Hegemann, a high-school student from Newman Grove, Nebraska, proudly displaying a scale model of his Pivot Track Eliminator, a tillage device. Leonard Berghoefer of Hampton, Iowa, sat quietly next to his Calf Tote, a human-powered buggy for moving newborn calves. Arvin DeCook, a young farmer from Sully, Iowa, displayed a simple backhoe implement he had made from scraps of steel. Kevin Urick of Prophetstown, Illinois, handed out information on his seed-spreader system with the zeal of John the Baptist. Urick wasn't interested in patenting his idea, and stood to make nothing from it. "I just think it's something that a lot of farmers would like to know about, and I'm excited to have this chance to share it with them," he said.

Most of the contestants, though, possessed at least modest commercial ambitions. Take, for example, Jerry and Robin Moore, a young farm couple from Winthrop, Iowa. Jerry called his idea the Feed Buggy. He had undergone hip-replacement surgery following an accident, which made it difficult for him to carry large containers of grain ration to feed his sows. To ease this painful chore, he designed and built a two-wheeled wagon that could be hitched to a lawn tractor. The Feed Buggy had a hydraulic auger that could be activated from the seat of the tractor. At feeding time, Jerry, Robin, or their twelve-year-old son would drive the lawn tractor and Feed Buggy slowly down the rows of concrete feed bunks, doling out the proper amount of ration to each sow. The Moores were quietly hoping the Feed Buggy would generate extra income to supplement their farm earnings, which they admitted were slim.

Robin, who was pretty, petite, and thirty-eight years old, spoke with disarming, and poignant, frankness. "We're barely hanging in there financially," she said. "We made it through the eighties somehow. I worked as a veterinary's assistant for a while, then quit to take care of our two kids. But things are getting tougher again. I'm thinking I'm going to have to leave the kids and go back to work."

I asked what kept them on the farm if things were so tough. Robin thought for a moment, then said, "Well, in the spring, when you go out and start working the land, it's a real rush. There's nothing like it.

"They say farming isn't a way of life, it's a business," she added. "But it's a way of life, too, because if you didn't have that, there'd be nothing else to keep you there."

One of the most popular Idea Exchange booths belonged to William Jones, a retired farmer from Tuscola, Illinois. His wife, who was there with him, had small hands and suffered from arthritis. For years she had found it nearly impossible to open sealed jars and bottles. So Jones scrounged around in his shop and made a mechanical jar opener using an old coal-stoker gear box, parts from a broken tractor clutch, a salvaged electric motor, and a couple of rat-tail files. His unscrewer worked, but it was too big and ugly to sit in the kitchen. Jones decided he had a marketable idea, so he built a more compact model and, over the next six years, spent $10,000 to patent and market his invention. Eventually he licensed it to the Appliance Science Corporation of Norwalk, Connecticut. Hammacher Schlemmer and other upscale retailers picked up Jones's creation, called the Open-Up Power-Twist Jar and Bottle Opener, and sold it for about $35.

Jones was the big daddy of the Idea Exchange bunch—a farmer-inventor who had played the odds, spent the money, found a manufacturer, and made it to the big time. I asked him what he thought of the two guys in the nearby booth who were trying to reinvent the combine. "Well," Jones said, "I've owned a combine for many years, so I understand the principles involved. I think their idea is basically sound, but they're going to have a hard time getting a major manufacturer to bite. With my idea, there were more than twenty different companies I could go to. A lot of companies make small appliances. But only two or three companies make combines, and combine sales are way down. The odds aren't in those guys' favor."

Ralph and Mark's booth was positioned at the end of a row, next to a wide aisle that carried a heavy flow of foot traffic. "One of the things the judges will be looking for," Ralph told me, "is how much attention each booth is getting from the crowd." He and Mark were getting a lot of attention. They had set up a bulletin board covered with pictures of Whitey, along with diagrams showing how the Bi-Rotor mechanism worked and charts comparing the performance of their prototype with that of a conventional combine. Mark's original model (the one he could operate with an electric drill) sat on a table under the bulletin board. Next to it, a big-screen TV monitor displayed a video. The ten-minute tape began with a flashy news segment produced by a TV station in Denver. It showed Mark driving Whitey through a field of wheat while a newscaster's voice said, "Mark Underwood is the kind of guy who gets things done—an Eli Whitney sort of a person."

On the first morning of the convention, Mark switched on the video and watched it for a minute. He blushed at the Eli Whitney reference, flattered by the comparison. "You know," he said, "Eli Whitney was a great inventor, but he got screwed when everybody started pirating the cotton gin." Scanning the room with exaggerated suspicion, he added, "Just make sure one of them big manufacturers doesn't come up and stab me in the back." You could tell he wasn't entirely kidding.

As Bartle Hall filled with conventioneers, Mark and Ralph pumped hands like political running mates. They answered questions and struck up conversations, drawing people into their exhibit. The reaction to it was mixed.

"What, for God's sake, is this?"

"You mean the cage revolves, too?"

"Hey, I saw you guys in a magazine!"

"When are you coming out with that thing?"

"Well, I'll be damned."

"If you guys are so smart, how come that thing's not painted John-Deere green?"

"The majors are going to try to steal it from you. They hire consultants just to help them break patents."

"You haven't sold it yet?"

"Can I pick canola seed with that thing?"

"You got any nibbles?"

"Well, good luck. It looks like a true innovation."

By shortly after noon, Mark was ready for a break. I strolled around with him to look at the trade-show exhibits. We decided to eat lunch by grazing on free food samples. The National Corn-Growers Association was passing out popcorn popped in corn oil. The Louisiana Crawfish and Rice Promotion Board was handing out little cups of crawdad jambalaya. The Missouri Department of Agriculture offered a whole buffet of Missouri-grown products, from roasted walnuts and soybeans to cheese and sausage on crackers. One of the biggest displays in the hall belonged to Philip Morris. A long line of people snaked up to the food-and-tobacco conglomerate's pavilion, where four hirelings handed out shopping bags full of products such as Honeycomb breakfast cereal, Jell-O, Maxwell House coffee, Crystal Light Instant Sugarless Beverage, Minute Rice, and red corduroy baseball caps sporting the Marlboro logo. The people handing out the bags said, "Philip Morris is the world's largest food company, through its Kraft General Foods divisions." They chanted this over and over, as if administering a sacrament.

Late that afternoon, after the exhibition hall had closed for the day, a Farm Bureau official came to the Farmer Idea Exchange area to announce the four finalists. In alphabetical order, they were Arvin DeCook, Low-Cost Backhoe; William Jones, Open-Up Bottle Opener; Jerry Moore, Feed Buggy; and Mark Underwood, Bi-Rotor Combine.

A COMBINE IS called a combine because it integrates, in a single machine, three distinct operations: reaping, threshing, and winnowing. The reaping function is accomplished by a wide platform mounted to the front of the combine called a header. Headers are removable, like giant sets of mechanical dentures, and are designed for various purposes. The header you use depends on the kind of crop you want to harvest. (A header for small grains, like wheat and oats, has a cutter bar and a rotating reel, much like a McCormick reaper; a corn head has no reel, but, instead, has fingerlike projections that glide through the rows, snapping ears off the stalks.) Mark Underwood had redesigned nearly every major component of the modern combine, but, he admitted, "I'm not trying to reinvent the header." He had, instead, reimagined the way a combine did the other two-thirds of its job—threshing and winnowing. His major innovation, therefore, traced its ancestry not to the scythe and the cradle but to the flail and the wind.

A flail, at its most basic, is a branch or stick used to beat kernels of grain from their husks. The poet Robert Burns called flailing "the most degrading of occupations." I tried my hand at it recently on some wheat I grew in a garden-sized plot, thinking it would be fun and interesting to make a few loaves of bread from homegrown ingredients. I cut the mature wheat with a scythe, raked it into heaps, and carried the heaps to the barn, where I scattered them on a sheet I had spread over the hard-packed earthen floor. Using two sticks of uneven length, I had made a flail similar to the implement a pioneer homesteader might have used. The longer stick was a broom handle; the shorter one was a clublike piece of hardwood. I had drilled a hole through one end of both sticks, then joined them together with a short loop of rawhide. To use a flail, you hold the longer stick (called the staff) and shake it in such a way as to make the heavy, shorter stick (called the swiple) swing around in an arc and strike the ground solidly, with its full length, once every revolution: *Whack. Whack.* After some practice, I could make the flail go around with a regular, allegro rhythm: *whack, whack, whack, whack, whack.* In this manner, I beat the living daylights out of my wheat stalks for several minutes. Then I cleared away the straw with a pitchfork and found hundreds of plump little wheat berries scattered on the sheet.

Another way to thresh grain is to let animals tread on it, thereby pressing or rubbing out the kernels with their hooves. Animal-treading, or a variation on it, is still the way a large portion of the world's farmers thresh their grain. A step beyond simple treading involves hitching the animals to a threshing sledge—a heavy slab of stone or wood with a roughened bottom. Apart from the simplest hand tools, threshing sledges are the oldest agricultural implements in recorded history. A crude likeness of a threshing sledge appears on a clay tablet that dates from around 3000 B.C., unearthed by archaeologists not long ago in southern Iraq.

The sledge depicted on that tablet is what the author of the Book of Isaiah had in mind a few thousand years later (in about 700 B.C.), when he wrote:

> *Behold, I will make of you a threshing sledge,*
> *new, sharp, and having teeth;*
> *you shall thresh the mountains and crush them,*
> *and you shall make the hills like chaff;*
> *you shall winnow them and the wind shall carry them away,*
> *and the tempest shall scatter them.*

The term for "threshing sled" in ancient Rome was *tribulum.* The tribulum was a heavy block of wood with sharpened stones or iron studs driven into the bottom. The action of a tribulum as it was dragged across scattered stalks of grain presented a vivid image of a scarifying ordeal—the ultimate (and original) form of tribulation.

Flails and sledges stood at the pinnacle of threshing technology for millennia. In the Western world, this began to change in the eighteenth century A.D., at about the same time and place that the idea of mechanical reaping began to attract the interest of English and Scottish inventors. In 1734 Michael Meinzies of East Lothian, Scotland, petitioned George II for a patent on a machine consisting of a series of mechanical flails. According to the patent description, Meinzies's machine could beat grain at a rate of 1,320 strokes per minute, or the equivalent of "as many as 33 men threshing briskly."

Public sentiment toward the earliest machines of this kind ran from mild suspicion to violent hostility. Jethro Tull, the English agricultural innovator, who died in 1741, was publicly denounced for being "wicked enough to construct a machine, which, by working a set of sticks, beat out the corn without manual labor." The idea of a mechanical thresher seemed wicked to peasant and landlord alike. Because threshing by hand was so laborious, the occupation of thresherman was a common one among the peasantry. Doing away with the thresherman's task would put a large class of laborers out of work, thereby disrupting the entire social order. Even people whose lives would not be directly affected by mechanical threshing devices saw something sinister in them. A machine that threshed with the mere turn of a crank seemed so *unnatural*. Pious upholders of the faith of Calvin, Knox, and Cromwell believed that threshing by machine flew in the face of God's intended order.

Machines that winnowed seemed even worse.

Winnowing was the next step after threshing. After the larger pieces of straw had been removed from a pile of threshed grain, the kernels were still heavily mixed with bits of husk, leaf, stalk, and dirt—debris that present-day combine engineers refer to as material other than grain, or M.O.G. Winnowing gets rid of the M.O.G.

American pioneers winnowed their grain in several different ways. Most of them relied on natural wind. One old-time method was to pour threshed grain back and forth between two baskets or buckets, allowing a stiff breeze to carry away more and more chaff with each pour, until the kernels were clean. Another method was to take a large cloth, spread the threshed grain over it, grasp the cloth at its corners and toss the grain in the air, over and over. This is what I did with the wheat I had threshed with my homemade flail. Gathering the grain on the old bedsheet on which I had threshed it, I took the bundle outside the barn. I enlisted my wife to help me hold the sheet by its corners and toss its contents into the air. Only a light breeze was blowing that day, so a cloud of M.O.G. hovered listlessly over the sheet. The wheat was only moderately clean after dozens of tosses. I could see why, at certain times of year, pioneer farmers waited for wind as anxiously as becalmed sailors did. "This is productive of great inconvenience," wrote one British farm historian in 1811. "In a long track of calm weather, the corn is often unavoidably spoiled for want of wind."

Still another method for cleaning grain, one that also served as a crude method of sorting it by grade, involved taking a shovelful of kernels and tossing them across a clean wooden floor. The resulting mess looked something like a teardrop, or a comet. The material that flew the farthest, forming the comet's head, consisted of the heaviest, plumpest kernels. The comet's middle section was made up of some chaff mixed in with lighter kernels, most of them shriveled or split. This inferior grain in the middle came to be known as "middlings." The wispiest material, consisting mostly of bits of straw and husk, flew least far. This debris formed the broad tail of the comet and was called "tailings." By extension, the word tailings came to describe the residue from any process of purification or extraction, especially mining.

Farmers in southern China used winnowing fans during the Han dynasty, which started in 206 B.C. Such machines were rare in Europe, however, until after A.D. 1700. Winnowing fans were slow to catch on, in part because they seemed to defy the laws of nature even more than

A typical fanning mill, which separated grain from chaff. Some eighteenth-century preachers condemned such machines, saying they created an unnatural "devil's wind." (American Society of Agricultural Engineers)

mechanical flails did. When James Meikle introduced a fanning mill (a kind of winnower) in Scotland around 1710, local clergymen labeled it "devil's wind." Using such a machine, the clergy said, amounted to "impiously thwarting the will of Divine Providence, by raising wind . . . by human art, instead of soliciting it by prayer."

By that time, however, Pandora's box was open. Machines for threshing and winnowing grain began to appear in the patent books with increasing regularity. The wheels of rural change, hand-turned for centuries, began to rotate under mechanical power.

The earliest threshing machines imitated either the flailing motion of the human arm or the treading action of animal hooves. Few of these machines, though, worked as well as the arm or hoof they tried to mimic.

A Scotsman named Leckie seems to have been the first to design a threshing machine based on a new principle: the use of a drum rotating inside a cylindrical grate. In 1758 Leckie invented a "rotary machine which consisted of a set of cross arms attached to a horizontal shaft and the whole thing enclosed in a cylindrical case."

About twenty years later, Andrew Meikle (the son of James Meikle, who introduced "devil's wind" to Scotland) watched a machine much like Leckie's being demonstrated. After a few minutes of operation, the machine blew up. Young Meikle was, nonetheless, enchanted with the idea of cylindrical threshing. He constructed a metal drum with four iron "scutchers," or beaters, attached to its circumference. The drum lay on its side and rotated, powered by wind or water, at about 200 revolutions per minute. A pair of rollers fed grain stalks, headfirst, toward the spinning drum. The protruding beaters grabbed the heads and swept them downward, rubbing them against a pressure-loaded "breasting" plate that "scutched" the kernels from their hulls. Meikle's 1788 patent for a "Machine for Separating Corn from Straw" provided for the addition of "a pair of fawners," or fans, below the cylinder, "so as to separate the corn from the chaff." The grain fell onto a "jogging screen," or "harp," that shook the straw to separate it from remaining kernels. (In later machines, up to the present, the counterparts of Meikle's "jogging screens" would be known as straw-walkers.) By 1805 Meikle was selling machines that not only threshed but also separated and winnowed: the world's first combines.

Early combines—short for "combined thresher-cleaners"—made a deep impression on nearly everyone who saw them. An aura of magic attended the spectacle of entire sheaves of grain being fed into one end of an oblong box while straw and chaff billowed out the other end and a rivulet of clean kernels spilled into a container at the bottom. The machines emitted a loud bellowing sound (one well-known brand was called "The Bull" because of the noise it made) and vibrated as though possessed by demons. They had to be staked securely to the ground so they wouldn't vibrate themselves to pieces or move around. Some shook so much, even when staked, that they dug into the dirt, earning for themselves the generic name "groundhog."

The first threshing machines were powered by hand cranks, although the cranks were difficult to turn. They demanded at least as much effort as a flail did, raising the question of why anyone would want to substitute one form of drudgery for another. This led to the invention of the horsepower, a device that enabled a horse, in effect, to turn a crank. The earliest and most common horsepowers were of the treadmill type. Treadmills relied on the horses' weight to work, so the amount of power they could deliver was limited. As threshing machines got bigger and their capacity increased, a new type of horsepower, called the sweep, came into widespread use. A sweep looked like a carnival pony ride. It consisted of a central spindle with long wooden beams, or sweeps, that radiated outward. Horses—from one to ten of them, depending on the machine's size and configuration—

Threshing with a horse-powered treadmill, c. 1850.
(U.S. Department of Agriculture)

were hitched to the beams. As the horses walked around in a circle, they turned the central spindle, setting a series of gears in motion that caused a long drive shaft to spin. This shaft, called the tumble rod, ran along the ground, a few inches above the dirt, for fifteen or twenty feet to the machine that needed power. The horses had to step delicately over the tumble rod each time they came around the circle.

The mechanical thresher, like the reaper, was born in Britain but grew up in America. Probably the first thresher imported to the United

States was one of Andrew Meikle's Scottish machines, in 1788. It took two people to operate (one to crank, one to feed the grain) and could thresh something like seventy bushels of wheat in a day—seven times as much as a person could thresh with a flail, or three times as much as two flailers could produce. American inventors seized on Meikle's basic principles and applied them to machines of ever-expanding size and capacity. By the late 1820s, farmers in eastern states could choose from among some 700 different makes and models of threshing machines. Most were designed for use with a horsepower, and several could thresh up to 300 bushels of oats a day (oat kernels separated from the husk more readily than wheat, so the daily output of wheat was somewhat lower). The machines were quite expensive, though. An 1822 advertisement, for example, listed a four-horse "combined thresher" at $500—an impossible outlay of cash for an ordinary farmer. Most buyers were wealthy farmers or custom threshers who took their machines from farm to farm for hire. By 1843 the *Richmond Planter* could report that "the use of the threshing machine is universal in Virginia." One of the earliest thresher manufacturers in that state was Robert McCormick, Cyrus's father. In upstate New York, the father of another inventor made his fortune in the threshing-machine and horsepower business. The name painted boldly on his machines was Westinghouse.

Pope, Westinghouse, Pitts, and J. I. Case were a few of the better-known American threshing-machine makers by the middle of the nineteenth century. Their earliest models were all of the groundhog type, staked down in one place. It was soon discovered that a pair of reciprocating sieves would keep the machines from vibrating so violently, allowing them to be mounted on wheels. These thresher-cleaners qualified as combines, but not yet in today's sense. Reaping was still a separate operation, performed either on foot with cradles or with one of the new, and still experimental, horse-drawn reaping machines.

Even as Cyrus McCormick and Obed Hussey were working out the bugs in their first reapers, a pioneer in the Michigan wilderness who knew nothing of their efforts did something quite remarkable. He built a machine that reaped, threshed, winnowed, and bagged grain, all at once—a supercombine that seemed to do everything but slop the hogs and milk the cows.

The story of the Michigan combine begins with John Haskall and his wife (whose first name the patriarchs of history have failed to record). Mr. Haskall was an attorney practicing in New York when, in 1820, he was named in an exposé of Freemasonry and decided to head for the frontier. Shortly after that, Mr. and Mrs. Haskall turned up in the small settlement of Kalamazoo. Lawyers were not in great demand in Kalamazoo at that time, so John Haskall looked for some other way to make a living. One day when he was traveling the wooded countryside near Kalamazoo, he came upon an utterly flat, treeless grassland encompassing at least 20,000 acres that was known as Prairie Ronde. He observed that some farmers had planted wheat on Prairie Ronde, with results that appeared quite excellent. If only he had a team of horses, he thought he could guide the plow well enough to till some of the fertile prairie and make a respectable living as a wheat farmer.

The main difficulty Haskall foresaw would come at harvesttime. He knew that even a seasoned farmer could not reap, thresh, and winnow scores, perhaps even hundreds, of acres of wheat by himself, and hired help was nearly impossible to find in Michigan. John Haskall thought of Prairie Ronde longingly, as if it were a succulent apple that dangled just beyond his reach.

Haskall confided his idea about raising wheat on Prairie Ronde to his wife. A night or two later, she had a dream. In her dream she saw many horses pulling a giant machine across an oceanic grassland. The machine was harvesting wheat. In some detail, she described to her husband how the machine worked. Haskall related these details to a mechanically minded farmer he knew named Hiram Moore. Moore lived near the village of Climax, on the edge of another expanse of wild tallgrass known as Climax Prairie. "But how would it cut the wheat?" Moore asked. Haskall held out one hand, fingers spread wide. He held the other hand perpendicularly on top of the first, sliding the top hand back and forth like a reciprocating saw blade. "It would cut like that," he said.

Hiram Moore had intended to let the matter drop, but Mrs. Haskall's dream kept intruding on his thoughts. Finally, after six months of this, he made some drawings and built a model. He sent the model to the patent office in Washington, D.C., and soon received a patent for a combined reaper-thresher. The year was 1834, the same year Cyrus McCormick got his first reaper patent.

By the following summer's harvest, Moore had made a prototype, minus the threshing apparatus. He took it out into a field and tried cutting wheat. Before he had gone forty feet, something broke. Even so, Moore felt encouraged. "I see the shore far off and it will take me a long time to get there," he said, "but I will succeed in time."

In the fall of 1836, Andrew Y. Moore (no relation to Hiram Moore) moved from Pottsville, Pennsylvania, to Michigan and planted a large acreage of wheat on Prairie Ronde. Hiram Moore asked his new neighbor if he would leave three acres of wheat standing so he could try out

The only surviving photo, taken in the early 1850s, of Hiram Moore's Michigan combine. (Smithsonian Institution)

the latest improvements on his harvesting machine. A. Y. Moore, who later helped select a site for the Michigan State Agricultural College (now Michigan State University), was quite taken by the harvester. Pulled by twelve horses, it cut, threshed, and cleaned several bags of wheat from his three-acre stand. Moore calculated the cost of harvesting with the machine and arrived at a figure of eighty-two cents per acre. A. Y. Moore later wrote:

> I inquired of my brothers, who were farmers, as to the cost of harvesting and threshing in the ordinary way of cradling, raking and binding, shocking, stacking, threshing, cleaning, etc., and upon strict calculation it costs in that old process $3.12 per acre. The contrast was so great that I took an active part in [the harvester's] future. . . . Before another harvest, [Hiram Moore] said to me: 'Mr. Moore, I can invent, but I can't drive the horses.' I replied that I would drive the horses and assist him, and I did so each year.

In 1841 Hiram Moore invented what he called the Angle Edge Sickle, a cutter bar with many triangular "sections" fastened to it. Two sides of each equilateral section were sharpened so that the sickle would cut on both the outward and return strokes (his previous sickle had been a flat bar with pitched, sawlike teeth that cut in only one direction). Apparently Moore came up with his improved design without having seen the nearly identical sickle, with serrated triangular sections, that Obed Hussey devised for his reaper at about the same time. Hussey is often credited as the inventor of the reciprocating sickle bar, still a central feature of every combine, hay cutter, and sickle-bar mower in use today. But Moore, it seems, also deserves a place in the genealogy of this idea.

Hiram Moore found investors and steadily improved his machine. By the mid-1840s, one of his harvesters was able to cut, thresh, clean, and bag more than 150 acres of wheat during a single short Michigan harvest season. James Fenimore Cooper, traveling through Michigan in 1847, observed one of Moore's harvesters in action. Cooper, obviously dazzled by what he saw, described it in *The Oak Openings* (1848). He gave a detailed explanation of how the machine worked, then concluded on a grandiloquent note: "As respects this ingenious machine, it remains only to say that it harvests, cleans, and *bags* from twenty to thirty acres of wheat in the course of a single summer's day! Altogether it is a gigantic invention, well adapted to meet the necessities of a gigantic country."

People made special trips to Prairie Ronde just to see Hiram Moore's "gigantic invention." Moore was happy to oblige them. His favorite routine was to hitch up the combine (the sixteen-horse hitch was a spectacle in itself), cut and thresh a bag of wheat, rush the wheat to a nearby mill, bring the flour home, and make a batch of biscuits with it—all in one morning. At lunchtime he would serve the warm biscuits at long tables set out on the lawn.

AT THE FARM Bureau Convention in Kansas City, Ralph Lagergren demonstrated how his cousin's Bi-Rotor threshing device worked to anyone who would listen. He used his hands as animated visual aids, spinning them in circles like rotors, cradling them like grates, or tracing arcs through the air like flying kernels of grain.

"The basic difference between the Bi-Rotor and conventional

combines is that we have an active grate," he would say. "A normal combine has a stationary grate. It only covers about 120 degrees at the bottom of the cylinder. Since the threshing area is limited, the holes in the grates have to be fairly big, or the crop would just collect up on top of them. A lot of trash gets through those holes along with the grain, so you have to have an elaborate cleaning apparatus in the back end.

"With the Bi-Rotor," he continued, hands now flying into action, "the grate moves. We've got three-hundred-and-sixty-degree threshing, around the whole rotor. And since the grate is rotating in the same direction as the rotor, it intercepts the threshed kernels as the rotor flings them off. With three times as much threshing circumference, plus the active grate, you've got higher efficiency. You can accomplish the same or better results as a standard combine, but in half the space. A standard combine cylinder is nine to twelve feet long. Ours is four feet long. And the holes in the cage are smaller than the ones in a standard grate, so less trash gets through. You don't need all that cleaning apparatus in the rear end of the combine. We can eliminate a lot of weight, and have a more compact machine with half the moving parts."

"That sounds good," one of his listeners said. "When's it going to be on the market?"

"We're trying. We've got our foot in some doors. Now we've got to see who invites us inside."

A PHALANX OF metal detectors bisected the lobby of the Kansas City Municipal Auditorium. On the street side of the security barrier, conventioneers pooled and eddied like water behind a dam. They stood patiently, dressed in suits and Sunday dresses and pressed jeans with blazers and string ties. Most of them were farmers—farm men and farm women—and they had a scrubbed, red-cheeked look about them as they waited. The queues snaked out the doors and into the icy January morning. Secret Service agents ringed the lobby, scanning the crowd, while network television crews passed through their own security checkpoint. A mood of polite excitement swelled in the Art Deco auditorium as it filled with 5,000 farmers.

Mark Underwood and Ralph Lagergren stood in line, waiting their turn at the metal detectors. Mark wore no coat over his pinstripe suit, but seemed oblivious to the chill in the lobby. He surveyed the crowd solemnly.

"Mark's all pumped-up just to be able to do something like this," Ralph said. He looked more than usually pumped-up himself. "I've got to sit back with the crowd," Ralph added, "but Mark's sitting up front with the VIPs. Whoever wins the contest has to get up and give an acceptance speech." He nudged Mark in the ribs. Mark ignored his cousin. He smoothed his suit coat with his hands and peered this way and that, like a Secret Service agent.

I had journalist's credentials, so I had to go through the separate security gauntlet, where TV cameras, lighting equipment, tape recorders, and even my small notebook were all being poked and prodded and searched for hidden signs of felonious intent. Once my notebook and I had been cleared, I walked out onto the main floor of the auditorium in search of Mark. He was sitting up near the stage, in the second row of seats, immediately behind the Farm Bureau's top brass and in front of the governor of Missouri. I made my way to a seat in a block reserved for the governor's staff and sat down. Mark was seated with the three other Farmer Idea Exchange finalists. Sitting beside them was a fifth man, an executive from the Ford New Holland company, who would be announcing the grand prize.

President Bush's speech was what the crowd had come to hear, but that desire would not be fulfilled soon. The Farm Bureau had its annual meeting to run, and it would be anything but brief. First, two cinema-size screens above the stage carried an ag news report, followed by a specially produced video on farmers' property rights. Next came the parade of states: While an organist played "Three Cheers for the Red, White, and Blue," Farm Bureau presidents from all fifty states and Puerto Rico marched up the central aisle carrying flags from their home capitals. The marching presidents assembled on the stage and the meeting was officially brought to order with every pious and patriotic ritual known to man: convocation prayer, presentation of the flag, Pledge of Allegiance, singing of the national anthem—the works. The state Farm Bureau presidents were individually introduced, membership statistics from their states were recited, past presidents and national board members were recognized and presented with plaques.

Reporters in the press gallery, with deadlines looming for stories on President Bush's speech, grew fidgety.

Then came the speeches. *Everybody* gave a speech—everybody except for President Bush, who was nowhere to be seen. Dean Kleckner, the Farm Bureau's national president, delivered a seething address in which he lambasted environmentalists for questioning farmers' sovereign right to do whatever they pleased on their land. "Those who think a farmer's only responsibility is to look quaint for urban tourists on weekends don't like the way you and I are working our land," he said. "Sometimes I believe they don't like the fact that we *own* land. . . . Proponents of national land-use controls think the public interest is more important than our ownership rights. . . . Those who want to tell us how to farm without having the slightest idea of how food is produced are tenacious. But so is Farm Bureau." He was roundly applauded.

After several more speeches and a "stretch break," there was a stir in front of the stage. Men wearing trench coats and earplugs appeared and stalked back and forth, speaking into walkie-talkies. A man sitting next to me said, "I wonder what those guys have under their raincoats? It must be bigger than a pistol." Clearly, something important was about to happen.

And it did. The Ford New Holland executive stepped to the podium and said, "Ladies and gentlemen, the time has come to announce the winning innovative idea in the Farmer Idea Exchange competition. First, the runners-up. They are: Jerry Moore, Feed Buggy; William Jones, Open-Up Power-Twist Jar and Bottle Opener; and Arvin DeCook, Low-Cost Backhoe. The winner is Mark Underwood, Bi-Rotor Combine." Mark's head appeared, ten feet high, on the giant video monitors above the stage. His face, already flushed from the aerobic stretch break, was turning visibly redder. He was invited to the podium to say a few words. In a wavering voice, Mark thanked the Farm Bureau and Ford New Holland for sponsoring the contest. "My goal," he said, "is to get a new machine out there for farmers to benefit from in the future." It was the shortest speech of the day. Mark sat down and, a few moments later, a small Marine band struck up "Hail to the Chief."

PANNING FOR GOLD

Combining is a lot like panning for gold.

—Mark Underwood, while operating Whitey (1991)

It used to be said that you had to know what was happening in America because it gave us a glimpse of our future. Today, the rest of America, and after that Europe, had better heed what happens in California, for it already reveals the type of civilisation that is in store for all of us.

—Alistair Cooke, *Talk About America* (1968)

BY MAY 20, 1848, Edward Kemble was fed up. During the preceding two weeks the nineteen-year-old editor of the San Francisco *Star* had seen his readership all but vanish like a puddle in the desert sun. At least two-thirds of the town's male population had abruptly left their wives and children, abandoned their homes, and even deserted their military posts—all to pursue what Kemble knew to be a delusion. Only a month before, he had gone to John Sutter's sawmill on the south fork of the American River to check out the reports for himself. He had examined the very spot in the sawmill's tailrace where a carpenter named James Marshall sighted bits of a brass-colored mineral that he claimed was gold. When Kemble arrived on the scene, he had found no sign of gold whatsoever—no glittering dust, no euphoric miners, not a pan or a pickax. All he had seen was an unfinished sawmill, a glum-looking Marshall, a rather befuddled Sutter, and a Yalesumni Indian chief, who, by the campfire that night, had warned everyone that gold was bad medicine. At the top of his reporter's notebook, in large letters, Kemble had written, "HUMBUG."

Now, with the readership of his paper declining daily, Kemble published his opinion on the matter. The reports of gold were "all sham," he wrote—"as superb a take-in as was ever got up to guzzle the gullible."

Nine days later, Kemble's rival weekly, the San Francisco *Californian,* suspended publication for lack of readers. "The whole country," the *Californian* complained in its last issue, "resounds with the sordid cry of '*gold!* Gold!! GOLD!!!' while the field is left half planted, the house half built, and everything neglected but the manufacture of shovels and pickaxes."

Two weeks later Kemble gave up, too. "We have done," he wrote in the *Star*'s final issue. "Let our word of parting be, *Hasta luego.*" And then, having nothing better to do, he headed for the gold mines.

THE DISCOVERY OF gold at Sutter's mill in 1848 was a defining moment in American history. Its consequences rippled outward like the waves from a pebble tossed into a pond, ramifying in ways that were visible and invisible, immediate and long in coming, predictable and surprising. One of the most instant and obvious effects of the rush for gold was a local shortage of food. California agriculture, such as it was in 1848, consisted of a few large cattle ranches scattered up and down the grass-covered Central Valley. From unfenced herds of cattle, rancheros produced tallow and stiff hides, which the sailors who arrived to ship them away called "California banknotes." Spanish-speaking *Californios* grew fruit and vegetables on small cultivated plots, selling their modest excess in open markets like the one along San Francisco's embarcadero. The residents of San Francisco were obliged to obtain much of their food, including wheat for flour, from the states far to the east. Wheat grown in New York, Michigan, and Ohio made its way to eastern ports on wagons, trains, and canal boats. Then it was loaded onto oceangoing ships for the long voyage to San Francisco by way of Cape Horn—a distance of some 15,000 nautical miles, or five times the distance between New York and Glasgow.

Rumors of James Marshall's glittering discovery took a few months to reach the Atlantic seaboard. More months passed before the first argonauts arrived in California. But the onslaught of humanity, when it finally arrived, was overwhelming. California's non-Indian population, at the end of 1848, stood at around 14,000. By the close of

1849 it had grown to just under 100,000. Another 50,000 arrived in 1850. Late in 1852 the population of California approached a quarter of a million—an increase of 1,500 percent in four years.

Predictably, food prices shot up. Fortune seekers who arrived in 1849 found that meat sold for $3 a pound, potatoes and onions for $5 a peck. (Back East, $5 was a decent week's pay.) An egg in San Francisco cost $1. A live cow would bring $500, a small fortune.

To supply the urgent and explosive demand for food, farms sprang up like boomtown bordellos. On the outskirts of the settlement that became Sacramento, four men planted sixteen acres of potatoes and cleared $40,000 on the crop. Near Mission San Jose, a 150-acre vegetable plot netted $175,000 in one year. To make money on wheat, a relatively bulky and inexpensive staple, one needed more acres. More acres was precisely what the rancheros had. In the 1820s and '30s, Mexican authorities granted land to favored petitioners in large parcels. The parcels were measured by the league (one league is 4,438 acres); grants of three or four leagues were considered small. The Swiss immigrant Johann Augustus Sutter enjoyed one of the larger land grants of the Mexican era. His rancho near the confluence of the Sacramento and American rivers consisted of eleven leagues, or more than seventy-six square miles. Sutter was already growing wheat quite successfully before the gold rush interrupted his plan to found a colony, New Helvetia. More and more rancheros experimented with wheat as the demand for breadstuffs exploded after 1849. Their trial plantings thrived. As placer gold became increasingly scarce after 1850, many disenchanted fortune-seekers turned their hopes from mining fields to wheat fields, and the California gold boom gave way to a grain boom.

Compared with life in the California mining camps, the life of a farmhand looked good to James Patterson. He was a young argonaut from a farm in upstate New York. After trying his luck in the Sierra foothills with unspectacular success, he decided to return to something closer to the life he had known back home. In 1852 he went to work on the hacienda of John M. Horner, near Mission San Jose. Patterson was shocked to see Mexican Indians crawling through fields on their knees, harvesting grass and grain with sickles. The young man bought a cradle scythe, which enabled him to cut four times as much grain in a day—and earn four times the wage—as did the Mexicans with their sickles. Patterson knew, though, that even the cradle was old-fashioned, especially in wheat fields as big as the ones he saw in California. Where Patterson came from, farmers with any acreage at all used mechanical reapers to harvest grain. They threshed it with big threshing machines, powered by horses or steam engines. A friend of Patterson's, who worked at the famous Pitts thresher works in Buffalo, New York, had told Patterson how the mechanical threshers operated. While he was out in the big, swampy fields swinging his cradle, Patterson designed, in his head, a "traveling thresher"—a large, splendid, horse-drawn machine that would both cut *and* thresh grain. He described his idea to John Horner, and also to one of Horner's neighbors. The combined harvester-thresher sounded ambitious but workable. Anything that might give one wheat farmer an advantage in speed or labor cost over his competitors was worth trying, so the two farmers agreed to fund Patterson as he set out to build one of his machines.

Working feverishly for several months, Patterson managed to finish his invention by the harvest of 1853. The thing was colossal: Its wheels were eighteen feet high, and its threshing cylinder, at operating speed, let out a ferocious bellow that could be heard for miles. Twenty-two mules were required to move the machine forward and propel its ponderous internal apparatus. On its first day in the field, the noise it made spooked the mules. They bolted and dragged Patterson's machine uncontrollably across the countryside, breaking it to pieces.

In spite of the debacle, Patterson's ingenuity inspired John Horner. The idea of harvesting and threshing grain with a single machine appealed to him. The wheat business had suddenly grown quite competitive. On the strength of high prices, good weather, and good soil, the acreage devoted to wheat in central California had expanded geometrically every year since 1849. By 1853 the state was already self-sufficient in breadstuffs and even faced the prospect of an oversupply. Prices were leveling off. Efficient production, Horner saw, would be the key to the future. A machine along the lines of Patterson's could save him a lot of money in labor costs.

Coincidentally, Horner had heard of a man in Michigan who possessed a working machine that was much like Patterson's. The man with the combine was Andrew Y. Moore.

A decade earlier, A. Y. Moore had purchased one of the combines that Hiram Moore designed—the "gigantic invention" that James Fenimore Cooper had so admired. Cooper had predicted that such machines represented the future of American agriculture. In Michigan, however, they were duds. The big horse-drawn combines had failed to catch on for two main reasons. First, they were *too* gigantic and expensive to suit the scale of prairie farms, which were mostly small (less than 200 acres) and diversified. The average prairie farmer couldn't afford to keep the sixteen horses it took to run such a machine, let alone buy the machine itself, which would have cost thousands of dollars. A two- or three-horse McCormick reaper, still rather pricey at $150, was a far more practical investment.

The second obstacle was climate. Midwestern summers were hot and humid. Heavy rains could fall at any time during the growing season. These conditions produced wheat that remained high in moisture content until quite late in the season. At harvesttime, prairie wheat growers faced a dilemma. If they waited until the wheat was completely dry before reaping it, the reaper would dislodge many kernels from the brittle heads and spill them onto the ground. But if they reaped the crop while it was still slightly green and threshed it right away, the stalks would twist into ropelike bunches and clog the threshing cylinder. Reaping and threshing in the same operation was therefore an impractical idea in the humid Midwest. The best course was to reap the wheat when it was still slightly wet, then bundle it into shocks and leave it in the field to dry. One nineteenth-century authority on farm implements said that the Michigan combine would have met with more success in its native region if its inventor had also figured out a way to control the weather. Since he hadn't, A. Y. Moore's machine was one of the few working combines left.

The drier climate and bigger farms of California's Great Central Valley created a more receptive setting for the Michigan combine. John Horner recognized this and wrote to A. Y. Moore, suggesting that he send his combine to California to see how it would perform. Horner assured Moore that he and some neighboring ranch owners would pay Moore and a combine crew an attractive sum to harvest their crops for them. Moore thought this sounded like a good opportunity—speculative but potentially quite profitable. He shipped his machine from Michigan to an eastern seaport, where it was loaded onto a clipper ship bound for San Francisco. Moore himself stayed in Michigan but sent his son, Oliver, and a friend, George Leland, along with some good horses that were accustomed to pulling the machine. This small party of men and horses set out overland to meet the combine in San Francisco.

Machine, men, and horses all arrived safely, and they managed to combine more than 600 acres of wheat for Horner and two neighbors during the 1854 harvest season. Horner and Moore had been right: The combine was perfectly suited to conditions in California. The Michigan crew began to attract attention. One newspaper editor exclaimed:

> We saw it moving on its ponderous wheels. . . . At the close of the day's work, the harvester looks back and sees twenty acres of headless straw, while the decapitated grain lays over the broad field in well-filled bags, resembling hundreds of large sheep. This is one of the wonderful inventions of the age.

There was only one problem: California produced a bumper crop of wheat in 1854, and the local market could not absorb it all. Grain prices plummeted, and neither Horner nor his neighbors could pay the promised harvesting fee. Leland (who had purchased a half-interest in the combine) and Moore had spent close to $4,000 to get the machine and its crew to California. In spite of the combine's technical success, its owners had nothing to show for their investment. And unless grain prices rose, they never would. The Michigan crew put the machine in storage and, *faute de mieux,* joined the legions of miners.

Grain prices in California stayed depressingly low the following year, but by the summer of 1856 they had recovered somewhat. The crew from Michigan took the combine out of storage to try their fortunes again as custom harvesters. Unfortunately, though, nobody remembered to grease the machine. The crew hadn't gone far on its very first job when the smell of smoke filled the air. Friction had caused a moving part to heat up and ignite the highly flammable mixture of straw and chaff that lay thick on every wood and canvas surface. Soon the entire combine was engulfed in flames. Its crew managed to unhitch the teams and retreat to safety before wildfire blackened the whole wheat field. When A. Y. Moore heard the news, he decided he'd had enough of the adventure. "Thus ended the machinery business with me," he said years later.

John Horner had been party to a string of combine-related disasters, but his enthusiasm for gigantic harvesters remained undampened. He soon recovered from the near-bankruptcy that had prevented him from paying his harvesting bill. Not only did he resume farming on a large scale, but he became a combine impresario, building his own new machines and demonstrating them up and down the Central Valley.

Mechanically, Horner's combines bore a close resemblance to the Michigan prototype, with a few modifications. The most obvious change involved the placement of the horses. Horner put them behind the combine, where they pushed, instead of in front. Hitching the draft stock in front of the machine was more mechanically efficient, but this arrangement increased the likelihood of runaway horses and combine crack-ups. Another disadvantage of putting the teams up front was that hand laborers with sickles or scythes had to cut the first swath in a field, opening a path for the horses. (This was called "opening the field," and it had to be done for reaping machines, too.) John Horner's push-combines dispensed with this necessity and reduced the chance of stampedes. His first models measured twenty-two feet wide, twelve feet high, and thirty feet long. Much of this length consisted of a backward-projecting tongue made of heavy timber to which eighteen horses were hitched in a V-shaped formation. Horner built his first combine in 1859 at a cost of $12,000, which undoubtedly made it the most expensive farm machine on the planet.

A strong, nationwide demand for wheat after the Civil War enabled Horner to build two more of his combines. He named all three of them after the Union navy's ironclad ship the *Monitor.* Hoping to repeat the original *Monitor*'s victory, he set out with his fleet to conquer the Central Valley. His favorite strategy was to arrange demonstrations at some of the largest ranches in each wheat-growing district, publicizing the trials in advance with sensational handbills and newspaper notices. One of his typical advertisements (next page) ran in an August 1868 issue of the *California Farmer.*

Horner's claims apparently *were* well founded. The owners of some of California's biggest grain farms greeted Monitor Nos. 1, 2, and 3 by hiring Horner's crews to harvest their fields. Local farmworkers were less welcoming, since the combines were putting them out of work. They jeered and taunted Horner's crews as they traveled from farm to farm. Horner began receiving threats against his life and property. On a hot July night in 1869, someone set fire to Monitor No. 2, and it burned to the ground. A reward of $500 was offered for information leading to the arsonist's capture, but no one ever claimed the money.

Horner never fully recovered from the disappointment and financial loss he suffered that night (Monitor Nos. 2 and 3 had cost him

NOTICE!

There will be a public demonstration of the

Traveling Harvester,

MONITOR NO. 2,

Upon the farm of RICHARD THRELFALL, in Livermore Valley, Murray Township, Alameda County, on

Thursday, August 28, 1868

Commencing at 1 o'clock, P.M.

We claim that one-half of the expense of harvesting would be saved to the Farmer by using the Harvester; in fact, the entire expense of Threshing is saved.

Three men and twelve horses have Cut, Threshed, Cleaned, and Sacked, in good work-man-like manner, fifteen acres of grain per day—making five acres per man—a feat, we believe, never performed in America before! One and three-quarter acres to the man, working with the most approved machinery, is about the highest figure yet reached—one acre per man being nearer the general average.

Farmers! Come and see if our claims are well founded.

$10,000 each). A decade after the fire, he left California for good, not bothering to take the two surviving combines with him. He abandoned them in the shed where they had been made, in a little village called Washington Corners. Horner's son, Robert, later wrote that people came to Washington Corners "from as far back east as New York to take measurements. I suppose that the present-day harvesters are more or less styled after Father's machine." And they were.

James Fenimore Cooper had cast the combine as a metaphor for the boundless promise of the American West. So far, however, the perfect "gigantic invention" for "a gigantic country" had left nothing but frustration and ashes in its wake. First came Hiram Moore, who won neither money nor recognition for building the machine that had appeared to his neighbor's wife in a dream. Next came A. Y. Moore, whose efforts to export the Michigan combine to California went up in smoke; then James Patterson, who saw his handiwork disintegrate behind a herd of runaway mules; and finally John Horner, whose three Monitors reaped a harvest of debts and disenchantment.

This trail of misfortune would soon lead to success. First, however, a few other things had to fall into place. The combine was only one piece in a picture puzzle of a new kind of agriculture—an agriculture shaped by the new "American System" of mass production and industrial capitalism.

What the combine already had going for it in California was a highly consolidated structure of land ownership, with a few farms that sprawled over several thousand acres. (As of 1875, only forty-five families controlled four million acres of the state's land.) That meant there were a number of farms on which a machine as big and costly as a combine might conceivably be operated economically. But the economic equation relied, first of all, on finding a large, stable market for California wheat and, second, on developing an inexpensive means of transporting the grain to that market.

The market turned out to be Great Britain. In spite of British farmers' increasing use of machines like McCormick's reaper, the United Kingdom's agriculture couldn't keep pace with the needs of its growing population. The potato famine in Ireland, which reached its worst point only a few years before the discovery of gold in California, forced Parliament to repeal its protectionist Corn Laws, which opened the British market to foreign grain. France, Prussia, Russia, and even India became major suppliers to Britain, as did the American Middle West, via railroads and steamships. California sent some exploratory shipments to the United Kingdom too, as soon as farmers in the Central Valley began outproducing local demand in the 1850s. These shipments were well received in Liverpool. The arid climate of the Sacramento and San Joaquin valleys produced a wheat with exceptionally good milling and baking characteristics: The kernels were hard and dry, and the flour they yielded was unusually white. Bakers' and millers' agents who frequented the Liverpool Corn Exchange started paying premium prices for California wheat.

The problem now was transporting the wheat from California

farmers to the Liverpool market, where world grain prices were effectively determined. San Francisco lay practically half a world away from Liverpool (although, surprisingly, Liverpool was no farther from San Francisco than Boston or New York were by way of the Cape). Fuel for steamship voyages of such distances was too expensive, given the relatively low value of a grain cargo. Clipper ships had been invented in about 1850, in large part to serve the gold-rush trade, but grain was too heavy and bulky for the ultra-streamlined clippers. Aware of the need for a fast, economical vessel for long-distance grain transport, shipwrights in Maine came up with a heavier, beamier version of the clipper. These ships became known as Down Easters. A Down Easter could put away impressive quantities of grain in its holds, yet it was fast. Ordinary freighters took 150 to 180 days to get from San Francisco to Liverpool; a Down Easter could sprint the distance in only 100 days.

By the late 1860s all of the basic ingredients for a thriving California wheat trade existed: eager sellers in San Francisco, enthusiastic buyers in Liverpool, and ships that could transport grain from one to the other quickly and economically. Still, there was a logistical hitch. The demand for export tonnage at grain-shipping time far exceeded the need for import tonnage to San Francisco. As a result, there were never enough ships in the Bay area when grain farmers most needed them. But shipowners wouldn't send empty vessels all the way to San Francisco unless a profitable cargo was guaranteed to be waiting. Shipowners weren't willing to take the risk—there was too much uncertainty about how much tonnage wheat farmers would, or would not, need in a given season. So, for lack of shipping, California wheat either rotted on the piers or sold for a song in the saturated West Coast market.

To solve the problem, some kind of controlling intelligence was needed—an omniscient power that could anticipate the size of the California harvest, estimate the number of ships needed to carry it, charter the ships for arrival in San Francisco at the right time, and live with the financial risk involved. An individual functioning in this capacity would need not only a thorough understanding of the world grain trade but also the intestinal fortitude of a high-stakes gambler and the killer instincts of a shark.

This combination of traits came together in Isaac Friedlander. A child of German Jewish parents, Friedlander had immigrated to the United States as a young child. He grew up in South Carolina, then joined the gold rush to California in 1849. Like other savvy forty-niners, he soon realized that the surest way to make a fortune in California was not to pan for gold but to supply food and other goods to those who did. In 1852 Friedlander secured enough credit to corner the flour market in San Francisco, eventually selling out at an enormous profit. He lost this fortune a few years later in another flour battle, but he continued to play the market and quickly recovered his losses. With a few partners, he built one of the first flour mills on the West Coast and began dabbling in grain exports. He shipped some flour to China in 1854, then some wheat to Australia and Britain a few years later, making a killing on both deals.

A giant of a man—he stood six feet seven inches and weighed around 300 pounds—Friedlander became the Michael Milken of a new market: California's emerging export trade in breadstuffs. Soon he was being called "the grain king," and it wasn't entirely meant as a compliment.

The grain king was a born speculator. His most important insight was an early grasp of the importance of shipping in the cycle of growing, harvesting, and marketing grain. At the outset of each growing season he developed a careful, close estimate of the size of the grain harvest on individual farms up and down the Central Valley. Then he chartered the necessary number of ships, assuming a virtual monopoly over inbound shipping. Anybody who wanted export tonnage had to go to Friedlander and pay his asking price—and that price was usually exorbitant.

Friedlander's enterprise prospered, but it was not without risks. The main peril for him lay in the possibility of overchartering. A late-season drought could leave him with the obligation to pay for more tonnage than there was wheat to carry. If he miscalculated, his entire, elaborate financial edifice (which was built on enormous lines of credit) could come tumbling down. He avoided this by building up an unsurpassed network of intelligence sources that kept him informed not only of crop conditions all over California but also of shipping movements throughout the world. Transcontinental and transatlantic telegraph cables, both completed in the 1860s, aided him tremendously in his bold venture.

Farmers came to hate Friedlander. He grew ostentatiously wealthy off California grain, while farmers themselves often labored near the margins of profitability. As the farmers saw it, Friedlander was sucking away their profits. His riches consisted of money that should have wound up in *their* pockets. They saw Friedlander as the figurehead of a growing network of bankers, merchants, and middlemen who seemed to have farmers over a barrel. The *California Patron,* a newspaper of the agrarian Grange Movement, lamented that Friedlander "had the wheat growers so completely under his control (with the assistance of the merchants scattered throughout the State, who were his agents), that even with large crops, farmers were growing poor, year by year." Farmers complained bitterly that they simply could not get a decent price for their grain, and they blamed it on the grain king.

It is hard to know whether California farmers would have been better off without Friedlander. On the one hand, he was a kind of economic parasite, and he took an unseemly delight in wringing as much profit as possible from the California grain trade. On the other hand, Friedlander lubricated the wheels of commerce, facilitating trade on a scale that would have developed much more slowly, if at all, had farmers been left to their own devices. The grain king may have taken more than his share of the pie, but he helped make the pie bigger for everyone involved, including farmers.

Friedlander speculated in more than shipping. He owned a good deal of land and invested in California's first large irrigation canal. He was also one of the principal backers in building a grain elevator—the first such structure to appear on the West Coast.

Of all inventions, the steam-powered grain elevator was, in the words of historian William Cronon, "among the most important yet least acknowledged in the history of American agriculture." Elevators stood as immense monuments to the bulk handling and storage of grain, a development that had some interesting, even profound, implications.

For much of the nineteenth century, farmers who had any intention of selling their grain sewed it up in cloth sacks. Sacks provided convenient, inexpensive containers that took up little room when empty and, when full, could be moved by a single person. Buyers and sellers negotiated for particular sacks of grain from particular farms or locales. The quality of the merchandise naturally varied and was determined by direct sample. There was no abstraction: What you saw was what you got.

A sack-based method of handling grain worked well for waterborne transport. Because the water level of a river or bay often fluctuated, warehouses had to be built well above the water's highest normal level. The path from granary to boat varied and often crossed over levees, docks, gangplanks, and companionways. The easiest way to load and unload grain under such circumstances was by hand, in containers of manageable size and weight: sacks.

In the 1850s and '60s, as railroads carried more and more grain (especially to and from Chicago), these individual containers seemed to become more of an impediment than a help. Railroad tracks did not have to be separated from warehouses by wide levees. Lugging heavy sacks one by one began to seem rather inefficient and old-fashioned. How much easier it would be to load freight cars—and boats, too, for that matter—with *loose* grain by funneling it through chutes, like a liquid.

Elevators catered to this new way of handling grain. An elevator was simply a vertical warehouse. It differed from a standard warehouse in that it used steam-powered machinery instead of human stevedores to load and unload its storage compartments. There was great economic advantage in this less labor-intensive method. In Chicago,

Steam-powered grain elevators in Chicago helped put human stevedores out of work in St. Louis. (Scientific American, *Oct. 24, 1891*)

which quickly adopted elevators during the 1850s, moving a bushel of grain from a wagon to a freight car, or from a freight car to a lake vessel, cost half a cent. In St. Louis, where stevedores and sacks persisted, the same operation cost five cents. This disparity had predictable results. In 1850 St. Louis handled more than twice as much grain as Chicago, but only five years later, the situation was reversed. Chicago rapidly multiplied its lead after that.

On a tour of America in 1861 Anthony Trollope climbed to the top of an elevator in Chicago and was amazed by the sight. "I saw the grain running in rivers," he wrote, adding in wonderment that "these rivers of food run up hill as easily as they do down." Later in his description of the elevator he observed that "it was not as a storehouse that this great building was so remarkable, but as a channel or a river course for the flooding freshets of grain."

Mindful of what the combination of railroads and grain elevators had done for Chicago, Isaac Friedlander wanted to see a similar bulk-handling system develop in California. With his financial backing, the first elevator west of the Missouri River rose at Vallejo, on the narrow channel where the rivers of the Great Valley came together and emptied into San Pablo Bay, the northern lobe of the great bay that meets open sea at the Golden Gate.

Friedlander's elevator was as tall and domineering in its way as the grain king was in his. It towered above the channel like a mighty fortress. Even so, it never came to anything—at least not in Friedlander's lifetime. The enterprise failed because of a general reluctance in California to abandon the old sack-based system. Shipowners wanted to stick with sacks because they were afraid that loose grain would shift dangerously in cargo holds during the rough voyage around Cape Horn. Farmers, for their part, didn't like the idea of mixing their grain up with everyone else's.

The farmers had a point. Mixing grain presented problems. Farmer Jones might bring in a wagon filled with clean, dry, top-quality wheat, while Farmer Smith might present a load of wet middlings. Jones would naturally expect to be paid more for his wheat—but if you mixed it with Smith's, the good wheat would lose much of its value. Mixing, in other words, would obscure real and important differences between shipments of grain from various sources. Keeping individual lots of loose grain separate, on the other hand, would result in the loss of valuable storage space.

Chicago's solution to this problem was to adopt a grading system. When a delivery arrived at a grain elevator it was inspected and assigned a grade, or category. By 1860, in Chicago, there were ten different categories of wheat alone. These categories included three general types of wheat—red winter, white winter, and spring—which were further broken down into grades: club (the highest), No. 1, No. 2, and "Rejected" (for Farmer Smith's wet middlings). A farmer who brought grain to an elevator was issued a receipt indicating the number of bushels he had delivered and their grade. His grain was then mixed with other grain of the same type and quality.

The grading system was simple in practice but far-reaching in its consequences. "The development of the system of grading and of elevator receipts," as Henry Crosby Emery, a leading scholar of commodity markets, wrote in 1896, "is the most important step in the history of the grain trade." Why? Because it effected an alchemical change of a physical product of seed, sunlight, and earth into an abstract commodity of the marketplace. "The elevators effectively created a new form of money, secured not by gold but by grain," writes William Cronon in his book *Nature's Metropolis*. "Elevator receipts, as traded on the floor of [the Chicago Board of Trade], accomplished the transmutation of one of humanity's oldest foods, obscuring its physical identity and displacing it into the symbolic world of capital." This transmutation, as much as any machine, accounts for the development of American agriculture as we know it today.

In the old sack-based trade of St. Louis, farmers and millers met on the levee and haggled over actual bags of grain, which they could look at and sample directly. In the new world of bulk-storage elevators and standardized wheat grades, the traders who crowded the exchange floor in Chicago rarely, if ever, saw real grain. As the years went by, an increasing number of them had no interest in owning any wheat at all, even though the slips of paper they traded gave their bearers title to boxcars full of it. Sometimes the bearer really did want his boxcar loads, so he would redeem his elevator receipts for real wheat and ship it off somewhere. But most of the transactions on the floor of the exchange did not require the grain itself to change hands. It just sat there in elevators like gold in Fort Knox. Or it flowed in Trollopian rivers, from barges to elevators to boxcars and back again, as the flow of paper at the Board of Trade dictated. The flow of paper, however, took on a life of its own, linked by a web of economic and psychological abstractions to that separate flow, down by the docks and rail yards, of golden kernels.

The slips of paper that flowed around the exchange floor fluctuated in value with the price of grain in Chicago. That price was determined in large part by prices at other major nodes along the grain river—places like New York, Buffalo, and Cincinnati. Prices at these ports along the river of grain all tended to seek the same level. They followed one another up and down as fast as transactions between ports could be completed. In the days of sack-based trade the pace of transactions was relatively slow, since physical grain samples had to move from hand to hand and city to city before buyers could be sure of what they were getting. But if Chicago and Buffalo used the same grading system, buyers and sellers in those two cities could make transactions without the need for samples. A bushel of No. 2 spring wheat was the same, for all practical purposes, in both places. With a uniform grading system, the speed of market-to-market transactions was no longer limited to the rather sluggish pace of the physical river of grain; it was limited only by the rate of flow of a new river—a river of information.

Starting in the late 1840s, telegraph lines carried information about commodity prices from Chicago to Buffalo to New York City almost instantly. Leisurely transactions on trading-room floors suddenly took on a more frantic character. The biggest prizes went to buyers and sellers who had access to the latest information and acted on it quickly. A broker in Chicago (for example), upon seeing on the wire that wheat prices in New York were rising, might rush out onto the trading floor and buy receipts for, say, 10,000 bushels of Chicago club wheat. The going rate for Chicago club on the floor at that moment might have been, say, eighty cents a bushel. Fifteen minutes later, after the Chicago market had followed the New York market up to eighty-five cents a bushel, the broker could sell his receipts and pocket a quick $500 profit.

Telegraphs, railroads, fast oceangoing grain ships, elevators, grading systems—all of these developments combined to create an unprecedented fluidity and volume of grain-based transactions between distant cities. As a result, these separate markets began to act more and more like a single giant market that spanned not only the nation but the globe. Local factors affecting supply and demand—droughts, floods, frosts, and the like—no longer had much, if any, effect on the grain prices posted at local elevators. Conversely, global events—such as the Crimean War or the entry of Australia and New Zealand into world commodity markets—affected farmers in Illinois and Wisconsin as never before.

Many farmers naturally reacted by expanding their farms. It was their hedge against price fluctuations that, increasingly, they could neither control nor predict. If prices went down, then at least they had more bushels to sell, and maybe they could still pay their bills. If prices went up, then the bigger they were, the better. They could put some money in the bank to see them through a bad year—or they could expand some more.

In a country where hired hands were expensive and land was relatively cheap, expansion required buying machines that could cover more acres, but without additional labor. The race to substitute capital for labor was on. This race went to the farmers who adopted new, labor-saving technologies most readily. As long as demand for grain exceeded supply and prices remained strong, then the early adopters did well. They made enough money to pay for their new machines, and then some. Then by the time everybody else had bought one of the new machines (as many farmers felt compelled to do just to stay in the race), demand had fallen off, and so had commodity prices. Hence, the late adopters found it more difficult to pay off their investments in land and technology. They were the ones who tended to fall by the wayside as the trend toward fewer and larger farms gathered momentum. Jefferson's agrarian ideal seemed to be ever more elusive.

The grading system spawned another form of grain trade, this one even more abstract than the first: the buying and selling of futures contracts. Futures trading originated with a type of contract between buyers and sellers who were in different cities. Under the terms of the contract the seller would agree to deliver to the buyer a certain quantity of, say, No. 2 spring wheat at a specified date in the future and for a specified price. This was called a "to arrive" contract. The parties to such an agreement said to each other, in effect, "Okay, this grain isn't going to arrive for a while. I know I might be able to get a better price if I wait until the arrival date and take the going cash price then. On the other hand, the going rate might not be so good by then. I can live with the terms we're agreeing to here, and I'd rather lock in on this price than take my chances by waiting."

A secondary market developed for these contracts as it had for elevator receipts. Buyers and sellers of to-arrive contracts were not buying and selling grain per se; they were really betting on whether grain prices would go up or down, and by how much, at some future date. Here's an example offered by William Cronon in *Nature's Metropolis:*

> Imagine, for instance, that Jones sold Smith a futures contract for 10,000 bushels of No. 2 spring wheat at 70 cents a bushel, to be delivered at the end of June. If that grade was in fact selling for 68 cents a bushel on June 30, Jones could either purchase 10,000 bushels at the lower price and deliver the receipts to Smith or—more conveniently still—accept a cash payment of $200 from Smith to make up the difference between the contract price and the market price. Had the wheat cost 72 cents on June 30, on the other hand, Jones would have paid Smith the $200.
>
> In either case, Jones and Smith could complete their transaction without any grain ever changing hands.

The *Chicago Tribune* estimated the volume of Chicago's cash grain trade in 1875 at about $200 million. The volume of the *futures* trade, by comparison, was around $2 *billion*—ten times greater. The difference would grow even greater in decades to come. Speculation in the futures market caused occasional dips and spikes in the cash market, which struck some people as a case of the tail wagging the dog. Farmers, in particular, felt increasingly excluded from the mechanism that determined how much they were paid for their products. Their economic fate, it seemed to many of them, lay less in their own hands than in the hands of an urban elite—an effete crowd that made financial sport of the commodities that they, the farmers, sweated (and sometimes went broke) to bring forth from the soil.

The system of futures trading that developed in Chicago appealed to speculators everywhere. San Francisco's Produce Exchange adopted it in 1882, and the Liverpool Corn Exchange followed suit soon afterward. From 1886 to 1891, "No. 1 Californian" wheat was, in Liverpool, the only recognized basis for futures contracts. California wheat was, in other words, the gold standard of the global grain trade.

Isaac Friedlander would have been proud if he had lived to see the day. But disaster struck him in 1877. Because of a drought, the California wheat crop fell far short of expectations. Friedlander had, however, chartered enough shipping (on credit, as usual) to challenge the Royal Navy. Many of the ships he'd paid for, however, wound up leaving San Francisco empty. The grain king's financial house of cards came tumbling down, plunging him into bankruptcy. He died a year later.

Friedlander left a permanent mark on the global trade that he helped to create. He staked out new commercial opportunities along the pathway from the field to the plate. By taking enormous risks onto his own shoulders, he diminished the farmer's risk of producing a crop that couldn't find a market. He performed a service that allowed farmers to concentrate on farming. At the same time, he exercised his power over the marketplace with an arrogance and ruthlessness that caused farmers to detest him. He became a symbol of the many forces that made farming a precarious way to earn a living. As the minister who spoke at Friedlander's funeral put it, "he gathered the grain crops of California in the hollow of his hand"—just as his successors, companies like Cargill, Continental Grain, Bunge, and Louis-Dreyfus, would later gather the grain crops of the world in theirs.

IN THE HEYDAY of the California grain trade, which lasted from about 1860 to 1890, more than twenty combine manufacturing concerns sprang up along the West Coast. In 1880, Daniel Houser of Stockton opened the first factory in the world that did nothing but mass-produce combines. Houser's banker, whose name was L. U. Shippee, liked the prospects of the combine business so much that he bought the patents of several small manufacturers and built a big factory of his own not far from Houser's. For a while, Stockton was the combine capital of the world.

Often it was not big manufacturers like Houser and Shippee but individual farmers tinkering with their own machines who pioneered the most striking innovations. George Berry was one such farmer-inventor. Berry, originally from Missouri, grew wheat on 4,000 acres near Lindsay, in the Sacramento Valley. He was constantly experimenting with new and better ways to harvest his six square miles of wheat. Combines were the obvious solution to high labor costs, he realized, but the number of draft animals needed to propel the machines was ridiculous. A combine with a fourteen-foot cut required at least twenty mules to pull it; larger machines needed thirty or forty. The driver of such a rig had to sit on a crow's nest perched ten or fifteen feet above the ground in order to see in front of the first team, way out ahead.

Even with a huge number of horses, combining was hard on the animals. It was done in July and August, when the sun blazed with subtropical intensity. George Berry got tired of seeing his horses die of heat exhaustion. He had seen the new "steam traction engines"—trackless locomotives—that were just beginning to appear on some farms, mostly to pull plows and operate stationary threshing machines. Berry decided to take the obvious next step, which was to pull a combine behind one of the self-propelled engines.

Berry's horseless combine was an agglomeration of separate machines that he grafted together like Frankenstein's monster. The beast was fashioned out of a twenty-two-foot header, a twenty-six-horsepower steam traction engine (the term "tractor" had yet to be coined), a six-horsepower stationary steam engine, and a mechanical thresher-separator. The traction engine moved the whole contraption forward, while the smaller steam engine (a Westinghouse) provided power for the header, thresher, and separator. Both engines ran off a single boiler,

The world's first self-propelled combine, 1886.
(F. Hal Higgins Collection, Univ. of California Library, Davis)

which was fired with straw ejected from the separator. The machine thus generated its own fuel; or, as Berry said, it was "fuelled off the land."

Berry's machine took him five years to build. When he finished it in 1886, it worked so well that he decided to make it bigger. By 1888 he had nearly doubled the width of its cut, to forty feet. He also installed headlamps that year, adding yet another first to a machine that had already broken all records. It was the world's first self-propelled combine, the largest harvesting machine in existence and the only one that could operate at night. Berry pronounced himself quite pleased with it:

> The machine I built this season for my own use is as near perfection as a machine for that purpose can be made. . . . I have averaged this season about ninety-two acres per day. I cut in two days 230 acres. It does not take any more men than I used last year to handle it, and it does about twice the work.

Within a few years, steam combines had become a common sight in the Central Valley. Economics provided a strong incentive for the switch from horses to steam. According to one calculation, harvesting wheat with a horse-drawn combine cost $1.75 per acre in the 1890s, while a steam-powered rig could do the same job for twenty-five cents. It cost money to save money, though. Only farmers with thousands of acres to harvest could justify spending more than $5,000 for a steam combine.

Beyond the expense, steam-powered combines had other drawbacks: They posed a constant threat of fire from flying sparks; their boilers sometimes exploded without warning; and the machines were extremely heavy. Much of the Central Valley wheat land was swampy during parts of the year, and 20,000-pound steam engines had a way of digging themselves into stupendous mud holes. The so-called steam traction engines were powerful, all right, but on soft ground they easily lost traction, spinning their wheels in deepening ruts.

One solution to this problem was to make the wheels wider and more buoyant so they would float on top of wet earth instead of sinking into it. Two brothers named Holt pioneered this idea. The Holts, John and Benjamin, were wheelwrights in Stockton who, in 1885, applied their wheelmaking skill to horseless combines. They began mass-producing steam combines in the 1890s, and before long they were making machines that were even bigger than George Berry's.

The Holts' Big Betsy, for example, was a steam tractor designed to resist sinking even in the swampiest fields. Its two drive wheels were each eighteen feet wide. With the engine in between, the entire machine measured forty-five feet from side to side. In 1893 the Holts attached a separator and a fifty-foot-wide header to Big Betsy. With an American flag flying from a tall pole in the middle, this machine—Holt No. 574—looked less like an agricultural implement than a battleship.

The Holts recognized that wheels the size of grain silos were an impractical solution to the traction problem, so they came up with a better idea. They put Betsy's younger siblings on continuous crawler tracks, like those on present-day bulldozers. Holt steam crawlers, it was said, had two speeds: slow and damn slow. But they kept moving. Tracks gave a forty-horsepower Holt engine the pulling power of a sixty-horsepower engine on normal wheels. The crawler-tractors could go nearly anywhere—across mud, over ditches, up steep hills. In fact, Holt farm crawlers inspired the first military tanks, which the British built during World War I. The Holt concern eventually merged with another combine maker to form the Caterpillar Tractor Company, which would, in the new century, become the largest manufacturer of heavy construction machinery in the world. Caterpillar sold its combine division to John Deere in the 1930s; in the 1990s, however, Cat would renew its interest in harvesting technology.

ONE DAY IN 1891 perhaps the largest, most majestic ship ever seen in San Francisco sailed through the Golden Gate on its way to Liverpool. It was the new Down Easter *Shenandoah* (not to be confused with the infamous Confederate raider of that name), flying a full two acres of canvas on its four masts. The ship was loaded with the largest grain cargo a single vessel had ever borne—more than eleven *million* pounds of California wheat. The cargo was worth $175,000, which was, coincidentally, what the magnificent *Shenandoah* had cost to build.

The ship's record-breaking voyage was a final moment of glory for the California grain trade. By the last decade of the century, wheat acreage in the Central Valley had diminished sharply, and continued to shrink every year. New competitors for Britain's grain trade—agricultural upstarts like Canada, Argentina, and Australia—had started to flood the world market with bountiful crops. International grain prices fell to record lows and stayed there. California farmers found

Threshing with steam—a common sight in rural America until well into the twentieth century. (Smithsonian Institution)

new and more profitable uses for their land. Hacienda owners divided their giant wheat fields into smaller plots, irrigated them with an expanding network of canals and ditches, and planted the fields with fruits, vegetables, orchards, and vineyards—the crops that would later make California the top agricultural state in the nation.

Wheat all but disappeared from California for a reason that went deeper than declining grain prices. As was true of Virginia and other states on the opposite coast, crop after crop of wheat had exhausted the soil. Yields declined year by year. The application of fertilizer and crop rotation were still new ideas at the turn of the century, and such practices seemed too expensive and impractical for a crop as low in value as wheat. In California, as in many other parts of America, wheat served as a pioneer crop. It spread across virgin grassland and thrived for a time, then faded as more enduring and profitable uses for the land could be found. California wheat gave way to California produce, just as wheat in the midwestern tallgrass prairie relinquished its crown to a new monarch: King Corn.

THE INVENTOR

The American farmer has always grown ideas, as well as corn and potatoes. That is the secret of his prosperity.

—Herbert Casson, *The Romance of the Reaper*

The farmer and the artisan have more to do than they can perform; scarcity of men makes labor very dear; to supply the want of labor and time the American is forced to invent, to think out new ways of augmenting his efficiency.

—J. Hector St. John de Crevecoeur,
Sketches of Eighteenth-Century America (c. 1770)

If you want a thing done, don't give it to the man familiar with the art, who knows it cannot be done; give it to someone who does not know it cannot be done—and he will do it.

—Charles F. Kettering

If America now has about two million farmers living within its borders, then it must also have at least two million inventors. Farms breed inventiveness. It starts with the trait called "handiness," which boils down to a facility with tools and a wide-ranging understanding of how things work. Add a spark of imagination and you've got an inventor.

Some farmers invent out of practical need, others because they just can't help it. In spite of the thousands of farm tools and implements of every description available from manufacturers large and small, farmers often find a problem for which there is no ready-made solution. "There is a machine for every job on the farm," writes Verlyn Klinkenborg in his lyrical book *Making Hay,* "and yet much of the work, it seems, falls between machines." So the farmer fashions a gadget to do the job, using scraps and spare parts lying around the farm shop. Farmers make handy fence unrollers, sheep lifters, bale movers, and gate closers. They mount grain carts on rubber tracks, build customized tillage and spraying rigs, or assemble little chore buggies from scratch.

Such inventions are the exclusive subject matter of *Farm Show,* a fat, bimonthly tabloid that features 120 to 130 new ideas in every issue. "These guys are amazing," says Mark Newhall, the editor of *Farm Show,* speaking of some of the farmer-inventors his publication has covered. "A lot of them have better shops than the dealers who sell them their equipment.

"Wherever you go in the world," Newhall adds, "you don't find too many ideas coming from the engineering departments. They refine things. The basic new ideas come off the farms."

Frank Rowley's tornado machine is a case in point. Rowley farms near Wichita, Kansas, growing mostly wheat. A few years ago he was driving his combine home from a wheat field when a twister swept right over him. For a moment he felt the big combine shudder and dance weightlessly beneath him. The funnel passed on and found Rowley's son, who was driving a tractor hitched to a grain wagon. The tractor, the wagon, and Rowley's son were flung into a nearby ditch. They were a little battered and scraped, but otherwise not too much the worse for their ride.

Rowley was too engrossed in observing the tornado's behavior to be frightened by it. He was actually *glad* for the chance to observe a funnel at close range. For years he had been thinking about how to make an artificial tornado. The idea occurred to him one evening in the early 1970s while he was at home watching a television show. "At that time," he recalled recently, "everybody was terribly worried about how nuclear bombs were going to blow us all up. This TV show was trying to comfort people by showing us how a lot of *natural* forces are stronger than the atomic bomb."

Rowley didn't feel comforted. He was, however, fascinated by a segment of the show that dealt with tornadoes. The segment told of how a cyclonic vortex of winds, with rotational speeds exceeding 300 miles an hour, contains cells of varying pressure, from near-vacuums to several atmospheres. A tornado's destructive power, the show explained, comes not simply from the sheer force of the torquing winds but from the strong winds acting in tandem with the explosive variations in pressure within and surrounding the vortex. This "pressure gradient," as Rowley calls it, "explains how a tornado can tear down a house but leave a chair standing in the kitchen. After seeing that show, I thought, if I could capture the forces of a tornado in some way that I could control, just imagine what it could do. It would be one heck of a deal."

He had a particular purpose in mind for his captive tornado: He thought a controlled cyclone would make an excellent grain grinder. The idea arose quite by chance, as inventions often do. A friend of Rowley's who suffered from Hodgkin's disease had been placed on a diet of raw grain meal. The man's wife processed grain at home with a hand-operated grinder. One evening she called Rowley in great distress. She couldn't grind meal fast enough to feed her husband, she explained. The patient was losing strength, and she was losing heart. What should she do?

Rowley was mulling over this problem when he saw the TV show on nature's catastrophes. Why not capture the pressure gradient in a tornado, he thought, and use that to pulverize grain? It would cause much less wear and tear on machine parts, and maybe require significantly less energy, than a regular grinder.

Long after his neighbor recovered, Rowley continued to think about his tornado machine, mostly while he drove his tractor or did other farming chores. Finally, after sixteen years of intermittent noodling around on paper and tinkering in his shop, he felt ready to build a prototype. He thought he would need some help, though. A friend and fellow farmer, Pete Nusz, had some room in his shop, and he was a good mechanic. Rowley decided to share his idea with Nusz.

Later, Nusz recalled, "Frank called me and told me of this invention he wanted me to look at. We met at a McDonald's parking lot and got something to drink. We sat out in the vehicle until two a.m. kicking around the idea."

Rowley and Nusz formed a company they called Gradient Force and started working on the prototype. They finished it in six months. On an unseasonably warm February Monday not long ago, they demonstrated their prototype to a group of about fifty people, who had gathered in a parking lot at Wichita State University.

The machine was mounted on a double-axle trailer, which the men had towed to town with Nusz's pickup. On the trailer sat a metal framework supporting an elevated platform. Rowley stood up on the platform, a good ten feet above the ground. Near one end of the platform was a squirrel-cage blower powered by a forty-horsepower electric motor. At the other end stood a shining cone, fifteen feet tall. Its tapering lower half projected through a hole in the platform and came to a point just above the trailer bed. The cone resembled a small tornado sculpted from sheet metal. The blower and the cone, or "vortex chamber," were connected by a metal duct. On top of the duct, halfway between the blower and the vortex chamber, the machine's creators had mounted a small bin where objects could be introduced to the vortex.

All eyes were on Rowley as he flipped a switch. The blower accelerated to full speed, making a loud, whining roar, like a giant vacuum cleaner. Rowley picked up a bag of charcoal briquettes and poured them into the receiving bin above the platform. Within seconds, fine black powder began sifting out of the bottom of the vortex cone, where Pete Nusz stood waiting with a bucket. Next Rowley dumped in a box of sugar cubes, and out came fine sugar granules. Whole corn kernels, then soybeans, disappeared in the top and reappeared in Nusz's bucket as ground meal. Newspaper strips went in and emerged as envelope padding. Clumps of damp yard waste became small, dry flakes that resembled instant mashed potatoes. ("The machine dehydrates as it pulverizes," Rowley explained. "That's why we call it the Comminutor-Dehydrator. *Comminute* means grind or pulverize.")

For the grand finale, Rowley dropped some rocks into the feeder bin. The rocks were about the size of his fist. Apparently without ever touching the metal walls of the vortex chamber, the rocks spun around inside for a while, hurtling through pressure gradients and in and out of resonating frequencies, which caused the rocks to shatter like crystal when a soprano hits high C. "There's no mechanical grinding involved," Rowley said. "Except for the blower, the machine has no moving parts. The materials break up because of the pressure changes, combined with the resonating frequencies." Nusz proudly showed everyone the small shards of rock that had dropped out of the funnel into his bucket.

After the demonstration the small crowd retreated to a nearby building, where coffee and fresh cinammon rolls were served. A woman who ran a coffee shop in Rowley's hometown had baked the rolls using flour that Rowley had supplied. He had ground it from his own wheat using his pet tornado.

A group of aviation engineers at Wichita State studied Rowley's machine, hoping to figure out how it worked. They came away shrugging their shoulders, concluding mainly that its power requirement was surprisingly low.

"I don't have any problem with how it works," Rowley told me, "but the engineers can't explain it. Most engineers think if they haven't read it in a book, it can't be real. They don't understand vortex synergy. If you take that out of the equation, my machine doesn't work. But a bumblebee can't fly either, according to the laws of aerodynamics. Its body is too big and its wings are too small. It doesn't know any better, though, so it flies. Same with me, I guess. My mind hasn't been cluttered up with the nostrums of engineering school. I put big stuff in the top of the machine and get little stuff out the bottom, and I don't know any better than to think it works."

About thirty miles from the parking lot where Nusz and Rowley had demonstrated their tornado machine, Mark Underwood and Ralph Lagergren were preparing to build a new prototype of the Bi-Rotor combine. Their second prototype, Whitey's successor, would be a completely new machine from the ground up. "This is the step where we eliminate half the moving parts," Ralph told me. "It will put the Bi-Rotor threshing system together with several other innovative ideas Mark's been working on. When all his inventions come together in one machine, it will revolutionize the combine. This is our knockout punch."

The cousins wanted to build the new prototype in a secret location, away from prying eyes. Patent applications for some of Mark's ideas had not yet been completed, so Mark and Ralph thought it best to conceal their whereabouts from all but family and close associates.

They had negotiated for shop space with the owner of Kincaid Equipment Manufacturing, a small agricultural-equipment maker in Haven, Kansas. I drove to Haven one day in early March, when the fields of winter wheat in southeastern Kansas had already turned a St. Patrick's Day green. I knew I had found the right place when I spotted Ralph's pickup parked in front of a tan, metal-sided building on the south edge of town. Just as I pulled up next to the sporty red truck with the Texas plates, its owner appeared. He was wearing jeans, white tennis shoes, a gray Reebok sweatshirt, and a cap that said, "Parker Implement Co., Abilene."

"We've just been to a junkyard and bought ourselves a combine," Ralph said buoyantly. "It's an old International nine-fifteen. Mark figured it was better to salvage the running gear off of it instead of starting totally from scratch. We're pulling a Tucker."

Tucker—Preston Thomas Tucker—was a visionary inventor who, in the years following World War II, vowed to put the Big Three automakers out of business with his "car of the future." Detroit wound up putting Tucker out of business instead, but only after the inventor had built a prototype of his car using parts salvaged from junkyards. In 1988 George Lucas produced a film about him, *Tucker: The Man and His Dream,* directed by Francis Ford Coppola. Mark and Ralph loved the movie and often mentioned it. For them it was like a personal revelation, full of meaning, both inspirational and cautionary.

Inside the Kincaid shop, things were starting to wind down for the weekend. About a dozen men dressed in blue work uniforms busied themselves finishing various metal-fabricating projects. The space where they worked was a single, huge room. Scattered around the shop floor were four combines, three trucks of various sizes, two forklifts, and several large metalworking machines—breaks, lathes, a trip-hammer, punches, chain hoists, drills, mills, welders. Rows of steel racks two stories tall lined the walls. They were filled with miscellaneous hardware, raw steel stock, and subassemblies for Kincaid's products, which included specialized agricultural machines such as plot combines, field shellers for peanuts, and precision seeders. Parked in one corner of the shop, almost lost in the jungle of iron, was Mark's thirty-foot-long camping trailer. He and Ralph slept in the Winnebago at night, its porch light glowing softly in the dark, deserted shop.

Mark was working in a back corner of the room not far from the camper. He was invisible, lost inside Whitey's guts. I could see no sign of the new prototype, but various parts of Whitey were strewn everywhere. The inner rotor lay off to one side, and the outer rotor, or cage, sat near it on a wooden stand. The combine's feeder house, whose function was to collect the crop from the header and convey it into the main body of the combine, was detached and sitting near the combine's front wheels. Deep in the machine's thoracic cavity, I caught a glimpse of a blue-jeaned leg and a white running shoe. "Mark, are you in there?" I yelled.

"Hey, could you give me a hand?" said a voice from inside the machine. "Climb up inside the grain bin and hold these bolts so they don't spin while I tighten them up." I grabbed a pair of Vise-Grips from a toolbox, climbed up the combine's rear ladder to the engine platform, then crawled headfirst into the grain tank, where I saw some bolt heads wiggling down by the distribution auger.

"What are you doing to Whitey?" I managed to ask.

"She's our guinea pig," Mark replied. "I'm modifying her with a few more innovations. Mainly, we've been messing around with my new self-leveling sieves. Have you seen those yet?"

"Nope," I croaked. All the blood was rushing to my head.

"Well, come on down. Hey, Ralphie, are you out there?" Mark called.

"No, Ralph's on the phone raising money," said Ralph, standing below.

"Rig up that sieve switch so he can see 'em work."

I climbed down and ducked underneath the cowling at Whitey's rear end to examine the sieve panel. Mark extracted himself from the machine's belly and joined me. The focus of our attention was a big tray the size of a desktop, with perforations that could be adjusted in size for various crops. In operation, the sieve panel shook back and forth violently while air blew up through it. Chaff and straw were blown or shaken out the back of the combine, while the heavier grain kernels fell through the perforations and collected in a compartment below.

"The sieve is the combine's last defense against grain loss," Mark explained. "But it can't do its job efficiently unless the combine is level. The problem is, a lot of farm ground isn't level. On sidehills, chaff and grain kernels bunch up in channels on the sieve. The cleaning fan blows air up through the uncovered portion of the sieve, just escaping without doing any good. The grain and chaff don't get separated. The channels of grain and chaff just flow out the back end, like little rivers. There go your profits.

"The trick," he went on, "is to keep the sieve level, even when the combine is tilted sideways. Okay, Ralphie, jiggle the switch!"

As I watched, it became obvious that Whitey's big sieve panel was divided into long sections mounted side by side. Each panel, which measured about eight inches across, could pivot along its central axis, like the louvers in a window blind. When Ralph jiggled a small pendulum mounted on one side of the combine's abdomen, which acted as a gravity-controlled switch, the panels rotated from side to side. The gaps that opened up between sieve sections were covered with strips of flexible rubber. No matter how much the combine tilted as it moved across a hillside, the sieve sections would remain level.

"We tested the sieve earlier this winter at a friend's shop, up by home," Mark said. He told me they had simulated hillside conditions by tilting Whitey sideways with a hydraulic jack. Then, with the combine's threshing and cleaning systems running full tilt, they dumped wagonloads of unhusked corn into the mouth of the machine. A video camera mounted at the rear enabled them to watch how the threshed crop material behaved as it flowed across the sieve, with and without the self-leveling feature.

"Without the self-leveling device," Mark went on, "you could see the grain and chaff bunch up in rivers and channels when the combine was tilted sideways. The corn would just march right out the back end along with the trash. But with the self-leveling feature, the grain spread out evenly across the sieve. Ol' Whitey looked like she was about to tip over, but the material spread across the sieve as pretty as a carpet.

"It would probably have cost a major corporation a hundred thousand dollars to perform that test," Mark said, grinning. "We did it for practically nothing, with an old shop jack and a home video camera."

He looked at the sieve he had made. "The industry's response to the sidehill problem has been to level the whole combine body with big hydraulic cylinders," Mark said. "That's a thirty-thousand-dollar option. My solution is to level just the sieves themselves. The self-leveling sieve would probably cost about two thousand, and it could be retrofitted onto any combine. With it you can get twice the output on an eighteen-degree slope without slowing down. This sieve could save enough time and grain to pay for itself in a year or two."

After the sieve demonstration Mark hopped into the rear end of the combine, disappearing over the sieves with a tape measure. "I'm taking measurements for this new modification I want to try out," he shouted from somewhere inside Whitey's large intestine. "I want to put in some chaff-relief fans. If I could blow out the chaff from the rotor before it ever got to the sieve, boy, think what kind of capacity this old girl would have then! I've had this idea on my mind for a long time, but I didn't know what kind of fan configuration to use. But when I was walking through the junkyard today I saw an old apparatus lying on the ground, and it just clicked. Ain't that a hell of a way to get something done: Go to the junkyard for inspiration."

In the suite of offices connected to the shop I met two mechanical engineers who worked for Kincaid. They had had a little downtime lately, so Ralph and Mark were retaining them by the hour to help draft blueprints for the new prototype using computer-aided design, better known as a CAD system. Mark, however, was having a hard time committing his ideas to paper. He and the engineers would often spend a full day at a CAD terminal working on a design; then, that night in the camper, Mark would think of a refinement that changed everything. As I looked over the shoulder of one of the engineers, he brought up a combine design on the computer screen and proudly showed me how the graphic could be rotated to expose different views, or altered with a few simple strokes, or zoomed in to magnify a particular part.

"How have things been going with Mark?" I asked.

The engineer looked at me a bit glumly. After a pause he said, "We're on our sixth combine for those guys."

The inventor and the engineers, I soon learned, hadn't hit it off very well from the start. To Mark these engineers—almost any engineers—seemed plodding, uninspired, tethered by convention and conservatism. Mark harbored a deep fear that his baby would be "engineered to death." To the engineers, Mark seemed indecisive, ungrounded in technical training, and prickly. (In the end Mark decided to draw the blueprints himself by hand, and the engineers were dismissed.)

That evening Mark changed shirts inside the Winnebago before going out for supper. The interior of the trailer, with its brown faux-wood paneling and tan upholstered cushions, had the unkempt gestalt of a 1970s bachelor pad. A little plaque near the door said, "God Bless Our Camper."

"Ralphie and I are starting to put some things together—if we can just keep our confidence level up," Mark said as he buttoned his shirt. "People don't understand the direction we're trying to go in, so they get frustrated with us. But we're on track, now that we've got that International nine-fifteen chassis from the junkyard. We'll have to do some major alterations on the frame, but at least we've got something to start with."

Ralph got into his pickup and raced off toward Wichita. A single man now since his divorce two years earlier, Ralph was going out on a date. Mark and I drove into Hutchinson for dinner, settling on a Mexican restaurant where a line of melancholy mariachis performed near the cash register. Over platters of indistinguishable burritolike objects smothered in grated cheese, salsa, and refried beans, we talked about the roots of Mark's interest in inventing.

"I guess ever since I was little I've always thought mechanically," he said. "If I could see how to fix something, or make it work better, I was always tinkering with it. By the time I was eleven or twelve years old, I was tearing engines apart and putting them back together again. Mechanical things were my outlet for creativity. I souped up a Ford three-ninety engine while I was in high school. I bored out the cylinders, put in new pistons, added some gadgets of my own, and put it in a 'sixty-four Ford XL body. She was a pretty hot car, you might say. I tested 'er to the limits. I had confidence in what I had done. I pushed that engine into the red zone, and she never blew up on me."

While he had excelled at engine mechanics, the more conventional school fare—reading, writing, and arithmetic—had proved an unpleasant trial for Mark. He seemed to have a learning disability, but such problems were not widely understood in rural America during the 1960s. One teacher finally suggested that Mark might be dyslexic.

"They didn't know for sure what it was in those days," Mark explained, "but it was really a torment. Up through sixth grade I could hardly read a full sentence. I think I compensated for it with a strong visual memory. I could look at objects and see how all their parts fit together, and I could remember them. But that didn't help me with a lot of schoolwork. My mom worked with me every night for two or three hours, making me read my lessons out loud. I credit her for getting me through school. Without all the effort she put in, I probably would have dropped out of school and just worked on cars for the rest of my life."

Mark's mother, Wilma Underwood, brushed off Mark's tribute to her. "I don't know that I worked with him any more than most mothers work with their children," she told me. She went on to describe some of her youngest son's more dubious accomplishments.

"One of our neighbors once told me he'd be surprised if Mark reached manhood, because he was always inventing dangerous things—things he could blow himself up with," she recalled with a chuckle. "He would make his own firecrackers and light them off. When he was about fifteen years old he made a little metal cannon. It was real nice, with wheels and everything. It shot little metal balls and made quite a noise."

The relationship between Mark's presumed dyslexia and his keen visual intelligence is hard to discuss with any certainty. Some older theories of dyslexia link the disability to a lack of normal hemispherical dominance in the brain, which leads to an uncertainty about hand preference (Mark, as it happens, is ambidextrous) and a tendency to jumble the sounds and sequences of language. More recent research, however, has thrown this popular model of dyslexia into question. The term "dyslexia" itself, in fact, has fallen out of fashion, since investigators can't even seem to agree on a definition. Nobody knows for sure whether Mark's inventive talent might come, at least to some degree, from his efforts to compensate for his dyslexia or whether his strong visual imagination and his academic difficulties somehow go hand in hand. What does seem clear is that Mark's case is not without precedents. Edison was expelled when a schoolmaster branded him as "retarded." Elmer Sperry, another prolific inventor, was also an underachiever in school, in part because the standard curriculum failed to capture his interest.

No doubt Sperry, Edison, and Underwood would all have earned higher marks in school if the curriculum had been designed by Christopher Polhem, an eighteenth-century Swedish engineer and teacher. Polhem constructed a wooden "alphabet" consisting of basic mechanical devices—screws, gears, levers, cams, driveshafts, and the like. He told his students to compose mechanical "words" by assembling the "letters" in various combinations. His goal was to fill his students' imagination with these visual symbols so they would, in effect, become mechanical poets.

"The great independent inventors, such as Sperry, Edison, the Wrights, and Nikola Tesla, tended to be visual thinkers," Thomas P. Hughes, a professor at the University of Pennsylvania and one of the country's leading authorities on inventors, told me recently. "Sperry had a remarkable ability to visualize a machine and see it work in his mind's eye. Einstein once said that he found it very disconcerting to relate to people who were primarily verbal. I guess the lesson is that we shouldn't patronize visual thinking as being somehow less intellectual than verbal thinking. It's like your guy"—I had described Mark to him in some detail—"who obviously has this keen intellect of a visual kind. In many ways he sounds like a vintage independent American inventor."

After Mark graduated from Burr Oak High School he enrolled in the vocational agriculture program (better known as "vo-ag") at the community college in Beloit, Kansas. He concentrated on auto and diesel mechanics. The skills he learned there made him increasingly valuable at home on the farm. Gene, one of Mark's brothers, roomed with Mark at school and trained as an electrician. Oren, Mark's other brother, got a degree in animal science at Kansas State. Collectively the Underwood boys could fix just about anything, whether it had legs or wheels.

The brothers were accustomed to working together. For several years while they were growing up, they and their father journeyed to southern Kansas and hired themselves out as a custom-harvesting crew. They took a combine, a pickup, and a grain truck along with them, cutting and hauling grain for anyone who would pay them to do it. Working their way steadily northward as the wheat ripened, they eventually wound up back at home, where they would cut their own grain. Sometimes they kept going north, up into Nebraska. Mark usually operated the combine, while Gene and Oren drove the grain truck or helped their dad scout out work. Long hours in a hot, dusty cab didn't seem to bother Mark (this was before combines were air-conditioned), and he was always pushing the machine to the limits of its capacity. As a result, he could put more acres behind him in a day than most people could. He became intimately familiar with the machines in which he spent his days, developing a sixth sense about their care and feeding. "Mark was raised up with a combine," his mother said, "so he knows what they are."

The Bi-Rotor idea first occurred to him one day in 1971, during the summer before his senior year of high school. He had a summer job mixing cement for storm sewers using a drum-shaped mixer. "I was standing there watching the mixer rotate, mixing in the sand and gravel," Mark told me. "The idea just came to me. I suddenly visualized how you could thresh and separate grain, all in one operation, using that kind of rotating motion." He thought about the idea often as he plied the fields in a tractor or combine. He would be sitting there in a cocoon of engine noise—a kind of sensory-deprivation tank in reverse—and in his mind, slowly, methodically, he would review the plans for his new combine, adding to them and modifying them as if he were working at a drafting table with a pencil and an eraser. Then he would roll up the plans and stow them in a cubbyhole of his memory, where they stayed until the next time he pulled them out. For nine years he did this, telling only his wife, Deb, of the dream machine that was taking shape in his mind.

Mark had met Deb while he was going to school in Beloit, which was Deb's hometown. They were married in 1974 and lived, at first, in a little house in Burr Oak. Each morning Mark drove the few miles to his parents' farm. By the mid-seventies the three boys—Mark, Gene, and Oren—had their own fields to take care of (mostly rented land), in addition to helping their dad with the home place. They all helped one another out whenever necessary. They often worked as a team, especially during harvesttime. They moved from field to field with a couple of combines and grain trucks, eating up the acres just as they had during their custom-harvesting runs with their dad.

In 1980 Mark and Deb moved to his grandparents' old farm, just down the road from his parents' place. Mark was farming full-time on more than a thousand acres, while Deb took care of their two children (soon to be three). Not long after the move to their new place, Mark bought the two used bulldozers with which he started his excavation business. Mark enjoyed dirt work. It felt like sketching or drawing, but on a large scale and in three dimensions. Building long, gracefully curving terraces and waterways engaged the part of his brain that dealt with shapes, contours, and spatial relationships, which he found so much more congenial than words and numbers. He earned a reputation as the best dozer man in Jewell County.

That was during the 1980s—the worst decade for American farmers since the Depression. Dirt work provided Mark with a welcome reprieve from the stresses of home. The cockpit of a bulldozer, the cab of a tractor or combine, were his sanctuary, a place where he could be alone with his ideas. He and Ralph started pursuing the Bi-Rotor idea together in 1983. While Mark planted wheat and corn or built field terraces, he gradually reinvented the classic combine. By 1989 he had a vision of the ultimate machine. He and Ralph presented it, in the form of drawings and a model, to officials of the Kansas Technology Enterprise Corporation, better known as KTEC—a nonprofit corporation established by the Kansas legislature to incubate homegrown inventors and entrepreneurs. Much to the cousins' delight, KTEC awarded them a $20,000 grant for the purpose of testing the Bi-Rotor idea, thereby gathering data that would (they hoped) prove its validity. They took their grant and a testing proposal to the agricultural engineering department at Kansas State University. The chairman of the department was ambivalent about Mark's ideas, but he thought the grant money looked very interesting. He told Mark and Ralph, in effect, "Okay, we'll build a lab model and test it. It should take about a year and a half."

Mark and Ralph didn't feel they had a year and a half to spend waiting for faculty engineers to do the project in desultory academic fashion. They sensed that this was a make-or-break moment for their dream. This was the insight that led Ralph to quit his job. He and Mark went to K-State and took control of the testing project themselves. They camped out in a cheap hotel and, with the help of Sushil Dwyer, a doctoral candidate in the department, they built their lab prototype in twenty-seven days.

When they were finished they had a two-thirds-scale Bi-Rotor assembly: a whirling torpedo covered with rasp bars (the rotor), which rotated inside a funnel-shaped sleeve (the cage), which looked something like the cement mixer that had first inspired Mark, now shot full of holes. The whole unit was encased in a metal housing, with a drawer below to catch the threshed grain and a canvas bag at the small end of the cone to catch the discharged chaff. One cause of concern within the department was the question of how crop material would be fed into the Bi-Rotor unit. It was assumed that the testing team would have to build a precision-controlled conveyor belt, a task that could be as time-consuming and expensive as building the threshing unit itself. Mark and Sushil solved the problem by pressing a used, barnyard-variety grain elevator into duty. Adapting the paddle-and-belt-style elevator took a couple of days and cost practically nothing.

A test run went something like this: First the inner rotor was ramped up to a speed of around 800 revolutions per minute. Then the outer cage was adjusted to its proper rotational speed, which was usually a small fraction of the inner rotor's rpms. (One object of the testing was to try several different speed combinations.) The test apparatus whined and vibrated like a jet engine on the verge of takeoff. When the rotors were spinning at their test velocities, a signal was given to the conveyor operator, who flipped a switch. A dozen bushels of whole corn—husk, stalk, and all—went hurtling into the rotor assembly within the space of a few seconds. The sound was much like that made by a powerful wood chipper when someone throws a tree limb into it: *"EEEEEEEEOOOWWW!"* Then they shut the whole thing down, recorded data on rpms and threshing efficiency, and reloaded the corn conveyor for another trial.

Once they forgot to secure the grain-collection drawer tightly against the bottom of the rotor assembly. The box sat loosely at the bottom of its compartment, which left an open gap a few inches wide between the box and the threshing unit.

Mark activated the rotor, accelerated the cage to test speed, and signaled Sushil to feed the crop. Suddenly there was a sickening boom and the room instantly filled with chaff and dust. Ralph, Mark, and Sushil hit the deck, waiting for shrapnel from the machine to fly everywhere. But no shrapnel came. Mark shut everything off and waited a few moments for the dust to settle before inspecting the rotor unit for damage. Finding nothing wrong, he checked the grain box and saw that it was loose. Then he looked inside the box. It was filled with beautifully threshed grain, noticeably cleaner than it had been after previous trials.

"I knew right then that we'd made another important discovery," Mark recalled. "That's when I knew that not only did we have a completely new threshing mechanism, but we could combine the threshing and cleaning in one step. *Boom*—just like that—clean grain."

Data from the tests at Kansas State suggested that Mark's Bi-Rotor unit could do more threshing and cleaning in three or four feet than a conventional combine could accomplish in eight or ten feet. Armed with this data, Ralph raised some money, including another grant from KTEC, to fund the next step: building a full-size field prototype. The most expedient way to do this, the cousins decided, was to retrofit a Bi-Rotor unit inside the body of an existing combine.

They built Whitey at a machine shop near Dallas during the winter and spring of 1990–91. Fitting a Bi-Rotor into an International Harvester Axial-Flow combine presented some of the same challenges that one might face in grafting the guts of a jackrabbit into the body of a large dog. The Bi-Rotor assembly they fabricated measured four feet in length and about three feet in diameter at its mouth—about the size of a commercial clothes dryer. They had to fit this assembly into the space occupied by the combine's original rotor, which had the proportions of a nine-foot hotdog.

The operation qualified as major surgery. Mark was the head surgeon; Ralph served a dual role as assisting nurse and hospital development director. He convinced Goodyear to donate several thousand dollars' worth of tires in exchange for displaying the Goodyear logo on the new prototype, like an Indy racer. He persuaded Mike Logan, an autobody painter, and Jim Nelson, a mural painter, to donate their services. They imparted the finishing touches to Whitey after the surgery was over.

By that time it was early June, and much of the 1991 wheat crop in Texas had already been cut. Mark and Ralph had asked one of their investors, a farmer named Calvin Peterson, to leave a little wheat standing on his farm north of Fort Worth. Peterson had made out rather well when he sold some of his farmland to Ross Perot's son, Ross Jr., to become part of the Alliance Airport, the younger Perot's 16,000-acre industrial "superport."

In making fund-raising overtures to the Perots, Ralph had become friendly with several Perot employees, including a pilot named Dub Blessing. Ralph asked Blessing if he would go up in a Perot helicopter and take some aerial photos of Whitey on her maiden run. Blessing agreed. I now have a color Xerox copy of one of the resulting snapshots pinned to my bulletin board. Written across the photo with a black felt-tipped pen are the words: "Good luck Mark and Ralph on your Bi-Rotor Combine. Ross Perot."

On the day the photo was taken, the weather was clear with a few cumulus tufts scattered across a pale, humid sky. A huge hangar at the Perot airport dominates the scene, near the horizon. A little lower in the picture, a horizontal band of tan identifies Calvin Peterson's wheat field. Whitey shows up against the wheat as a light speck trailing a cloud of dust behind it. On her first test in real grain, she was working beautifully.

Now, nine months later, Mark and Ralph were camping out in yet another machine shop, working on another prototype. Mark had spent little time at home during the past three years. On the infrequent occasions when he did go home, he felt overwhelmed by the backlog of farm work and excavating jobs. He worked night and day, trying to catch up. He cut way back on his farm acreage and delegated most of the dirt work to his bulldozing partner, Von.

Still, he realized, his kids were growing up with a dad who wasn't really at home even when he was at the farm. There were always too many distractions. He thought about this as we sat in the Mexican restaurant in Hutchinson, with the mariachi band playing morosely in the background.

"There have been times when my kids have come back at me, 'Dad, you weren't at my ball game,' or 'Dad, you weren't at my track meet,' " he said. "Deb's doing an awful lot, keeping everything together at home, with the kids and all. I feel a little as if, here I am, pursuing my dream, but at their expense."

He paused, pushing a bite of burrito around on his plate. "I just hope I can do enough to soothe it over while I'm there. If we succeed with this, they'll reap some of the benefits, too."

The benefits he spoke of were monetary. Inventing holds—has always held—the allure of the lottery as a quick path to riches. But the odds of striking it rich by either method are extremely long, and Mark knew it. The privations he and Ralph had endured for three years—such as living like monks in a camper with no water or sewer hookup, using garbage bags for suitcases, eating meals at convenience stores—were things that the distant, uncertain prospect of money could hardly justify. There had to be more. Mark's greatest ambition, it seemed, was not material wealth but a success that would enable him to drive down Burr Oak's Main Street in his restored 1960 Sunliner convertible, the top down, smoking a cigar, secure in the knowledge that the people who once thought him slow in school were now aware that he had outsmarted the smart boys with their engineering degrees.

I asked him if he had any other inventions up his sleeve besides the combine.

"Oh, yeah, quite a few," he said. "Some are related to the combine—extensions of that idea. One or two are in the automotive area. Some others are totally different." He looked around the restaurant to make sure no one was listening, lowering his voice a notch. "I saw something when I was fourteen years old that I've thought about off and on ever since. There was an electrical storm one night. I was looking out my bedroom window and saw a bolt of lightning strike a tree. The lightning formed in a ball, and that lightning ball just floated slowly away from the tree. The energy in it defied the force of gravity. Someday I'd like to figure out how to harness that energy. Just like that old boy in Wichita who captured a tornado!" Mark had heard about Frank Rowley and his tornado machine. He grinned. "That's farmers for you. Always trying to fool Mother Nature."

Rowley, as it turned out, was having a hard time getting very far with his invention. He had retreated to a secluded shop to perfect the Comminutor-Dehydrator machine, and no one (including his wife) saw him or heard much from him for months on end. He seemed to be bogged down in the details of inventing, and without having a workable marketing plan. In short, he was a Mark without a Ralph. Unleashing his captive tornado would be all the harder for him because of it.

We returned to Haven at about eleven o'clock that night. Mark lingered outside the Kincaid shop for a while to smoke a cigarette. A sliver of new moon shone brightly in an almost cloudless sky, and there was a chill in the air. Mark looked up at the towering concrete silos of a grain elevator that stood across some railroad tracks. The smooth concrete silos seemed to phosphoresce a ghostly white. On top of the towers, a red warning signal blinked at the stars. Mark said, "You know, maybe what we're doing will help bring in the bounty of the world to fill these elevators. I'd like to think the amount of grain my combine and sieves will save in a year would fill these to the brim."

A BEEPING ALARM went off in the camper early the next morning. Since the trailer had no running water, Mark and Ralph were obliged to use the shop's bathroom facilities. The closest available thing to a shower was the fiberglass hand-washing basin on the shop floor, near a picnic table that the workers used for lunch and coffee breaks. A foot-activated valve on the doughnut-shaped basin caused water to sprinkle down from a column in the middle. To take a bath, Mark and Ralph wedged the sprinkler valve open with a block of wood, plugged the drain with paper towels, then mounted the basin, like King Triton in a fountain by Bernini. That morning it was Mark's turn to play Triton. By 6:15 he was splashing like a bird and yodeling loudly. Ralph, who hadn't returned to the shop until 3:00 a.m., was somewhat subdued as he got up and dressed for the day's activity.

Our plan was to drive up to Great Bend, Kansas, where the U.S. Custom Harvesters' spring convention was being held. Of all the many agricultural trade shows and fairs that take place in the United States each year, the Custom Harvesters' Convention attracts the most attention from combine manufacturers. That's because custom harvesters as a group provide the single largest market for new combines.

The 1,100 members of U.S. Custom Harvesters, Inc., range from farmers who supplement their incomes by doing a bit of custom work each year (as Delbert Underwood and his boys used to do) to big operators who own as many as eight or ten combines and do nothing but harvest grain for a living. The full-time harvest crews—"dedicated harvesters," they're called—start their season in southern Texas during the first part of May and slowly work their way north, reaching the Canadian border in September or October. Most of them work a wide band within the Plains states and do much of their traveling on north–south highways like U.S. 281, known in the business as "Custom-Cutter Alley." Each year in March the members get together to discuss issues of common concern, to see the latest equipment the industry has to offer, and to fortify themselves with fellowship—and maybe a beer or two—before beginning the long, lonely season ahead.

Great Bend sits at the apex of a big northward loop on the Arkansas River, right where the river intersects U.S. Highway 281. As we made the seventy-mile drive from Haven to Great Bend, we passed through the little town of Nickerson. "This is the place where Curt Baldwin started his first company," Mark said almost reverentially. "Baldwin was a true innovator in combines. I'd like to think we're part of that same tradition."

Curtis Baldwin was an independent inventor of the old school: a self-tutored, mechanically intuitive experimenter who was guided by practical experience and empirical observation rather than by formal schooling or theory. He was born in 1888 in a sod house in western Kansas and spent much of his boyhood working on migratory threshing crews, the antecedents of today's gypsy combiners. Baldwin started a custom threshing business of his own when he was barely into his teens. He spent most of his days pitching wheat sheaves into threshing machines, an activity that held no romance for him. "A machine that could make this drudgery unnecessary would be a blessing," he wrote years later, "and I was fired with ambition to build such a machine."

The machine he built, which would be the first of many, he called the Standing Grain Thresher. He obtained a patent for it when he was twenty-three years old. That year, 1911, he started the Baldwin Company in Nickerson, in partnership with his younger brothers, George and Ernest. Baldwin claimed that the horse-drawn (actually, horse-pushed) Standing Grain Thresher would reduce the cost of harvesting wheat from fourteen cents to two cents a bushel.

Curtis Baldwin was impetuous, independent to a degree that both friends and enemies cursed, and restlessly inventive. His mind churned with combines and harvesters. Around 1920 he became one of the first inventors to develop a tractor-mounted combine. He flitted from project to project, company to company, leaving mayhem, stepped-on toes, and new ideas in his wake. He claimed the trade name Gleaner for some of his machines, though his younger brothers had adopted the Gleaner name after Curtis left the Baldwin Company. (The Gleaner trademark still survives on one line of combines, though like most of the old names in agricultural equipment, it has changed hands repeatedly.) Curtis Baldwin eventually founded seven companies, every one of which failed. He hardly seemed to notice, though; he was too busy inventing. Toward the end of his life he was bent on developing machines that could harvest at higher and higher speeds. In 1955 he demonstrated a combine he called the Bearcat, racing it at fourteen miles an hour through a wheat field near Wichita. He died in 1960, dreaming of alfalfa-compressing machines and harvesters that worked on the principle of tornadoes. Like Preston Tucker, Curtis Baldwin was one of Mark Underwood's heroes.

The parking lot of the Holiday Inn in Great Bend was covered curb to curb with combines. Their names marched in bold letters across gleaming surfaces of sheet metal: Massey-Ferguson, Case International, John Deere, Gleaner. The machines were orange-red, blood-red, silver, and green, a distinct color combination for each manufacturer. Men wearing jeans and cowboy boots walked among the machines, kicking tires, climbing up into cabs. Many of them wore baseball jackets with the names of their custom-cutting outfits emblazoned on them. Robison Farms Custom Harvesting, Scandia, Kansas. Bryant Harvesting, Camargo, Oklahoma. Ehrlick Harvesting, Weldona, Colorado. The combine cowboys sauntered coolly with their thumbs hooked over their belts, toothpicks in their mouths, appraising the hardware.

Mark grew silently intense as he stalked among the machines, studying the newer features of each model. An hour or so after we arrived I found him perched on the upper platform of a Massey-Ferguson 8570, examining the hydrostatic unit that drove the threshing rotor. "You know, this Massey here has the smallest number of drive pulleys and belts of any combine on the market today," he said. "It has eleven. That Case-International over there has twenty-three. Our machine will have eight, maybe nine." His serious expression broke into a wide grin. "I can't wait till we get out with our machine. We're going to beat the pants off these guys."

I wandered across the parking lot to look at the Gleaner combines. The current owner of the Gleaner line was a new company called AGCO. Owned by a group of New York investors, AGCO seemed intent on becoming the Chrysler of the farm-equipment world, playing a nettlesome third to Case-IH's Ford and John Deere's GM. Standing in the shadow of a big Gleaner R-72, I struck up a conversation with Larry James, AGCO's friendly, soft-spoken chief combine engineer. James was widely known and respected as one of the top troubleshooting minds in the combine industry. I asked him if he had heard of Mark Underwood's Bi-Rotor idea.

"Oh, sure, everybody's heard about those guys," he replied. "I imagine everybody's watching their progress pretty closely. I haven't actually seen their prototype, but I'm not so sure why it's better than our system, or that it's much simpler. I remain open-minded." He looked up at the Gleaner that towered over us and shook his head, as though he were seeing not only the machine itself but the countless engineering man-hours and millions of dollars that had gone into making it.

After a pause he continued. "I'm interested. We're just going to wait and see. They have some major obstacles to overcome. You need to hook up with some major capital to make it all work. The tooling costs to set up and mass-produce something as big and complicated as a combine, that takes millions of dollars. Big companies like Deere and Case-IH have squads of accountants who say, 'The market is so many units. It will cost so much to develop and tool up to build it. How long is it going to take to make any money?' In normal times they look for a three-year payback. But times are lean now. Machines like this one"—he patted a side panel of the Gleaner—"aren't selling the way they used to. In today's market it would take a lot more than three years to pay back the cost of developing and tooling up for a radically different machine."

James, a lean, light-haired man, wore a black V-neck sweater and gray pants. A small crowd of interested custom-cutters had gathered to hear his quiet speech.

"The real question," he continued, "is, if Mark's machine is better, how much better is it? If it's only, say, five percent better than what we've got now, we would probably rather invest in our own existing technology and keep improving it. Even if their machine is twenty to thirty percent better, can you get twenty or thirty percent more profit from it? Can you make the kinds of return on capital that a big company expects? It used to be that the little guy with a good idea could set up shop, start his own company, and build machines. That's how all these farm-equipment companies got started. But it's not that simple anymore." Once again, James shook his head.

"There's another thing those guys have to watch out for," James said. He directed his attention kitty-corner across the parking lot, gesturing toward a fleet of machines painted the familiar green with yellow trim. "John Deere has a reputation, whether it's deserved or not, for flexing their muscles."

The Deere & Company display area was off to one side of the parking lot, a bit aloof from the other displays. Late in the afternoon I wandered over to the green machines with Mark. We climbed up into the cab of a John Deere 9500—one of the "Maximizer" series of combines that Deere introduced in 1989. The cab contained two seats, one for the operator and a smaller one for a passenger. Both were comfortably upholstered in earth tones. Mark settled into the driver's seat while I sat in the jump seat and closed the door. Suddenly we were enveloped in silence.

The panoramic curved-glass windshield was rimmed with switches, gauges, and digital readouts. One monitor provided data on engine

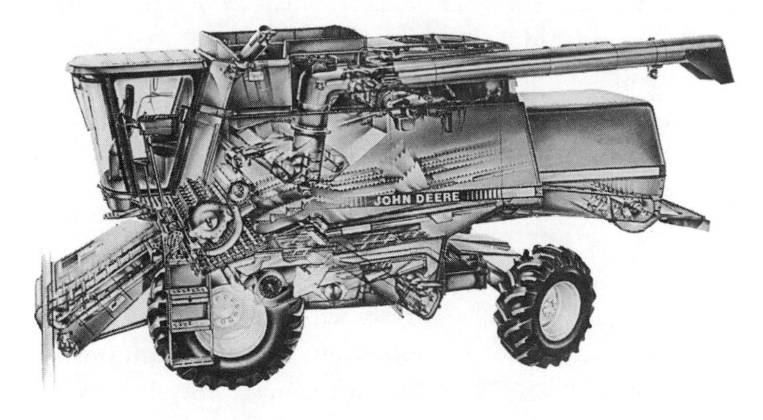

Cutaway view of a John Deere 9400 "Maximizer" combine, a farmer's factory on wheels. (Deere & Co.)

speed, cylinder rpm, cleaning-fan rpm, ground speed, and hours of operation. Overhead knobs controlled the stereo sound system, heater, and air-conditioner. A bank of red and orange lights could flash messages on nineteen different functions while the combine constantly monitored itself. At Mark's right hand there was an avionically inspired joystick called the hydrostatic ground control.

"It looks like you'd have to go to flight school to learn how to work this thing," I said.

Mark's mood was reflective. "You know, I've operated John Deere combines all my life," he said. "My dad had John Deeres, and my brothers still have green combines. They're good machines. But the company has gotten so big and full of engineers and executives, it's lost its ability to innovate. Take this cab, for example. It has the same basic controls and layout as the combines of twenty years ago, but they just keep adding more electronics, bells, and whistles. You ask somebody who owns one of these how he likes his new combine and he'll say, 'I sure like the cab.' But the thing that really counts is the machine the cab is sitting on. The guts of these Maximizers aren't that far removed from the threshing machines of seventy-five years ago. The old threshing rigs consisted of a cylinder, concave, beater, straw walkers, sieve, fan, and elevators. That's exactly what this one-hundred-and-forty-thousand-dollar machine has too, and in basically the same configuration.

"When they designed these new Maximizers they wanted to cut down on their grain losses," Mark went on. "So did they take some risks and try to figure out a more efficient threshing mechanism, to capture a higher percentage of the grain up front? No. They added another two feet on to the straw walkers. Instead of a real solution, they slapped on a Band-Aid."

After a while, I opened the door and we climbed down the ladder. On the ground Mark said, "Corporations provide the structure for manufacturing and marketing, but they can't innovate. That's the individual's role."

A SMALL SIGN ON the driveway leading to Deere & Company's Harvester Works Product Development building cautions visitors to watch out for browsing deer. The warning is no joke. Like all of the many John Deere facilities scattered in and around Moline, Illinois, the combine development building sits in a carefully manicured, parklike setting. Deer feeders on the grounds attract many wild corporate mascots.

From the parking lot the building looks like a windowless concrete bunker. But beyond the lobby, on the other side of a locked security door, the back of the building consists mainly of plate-glass windows, which frame outdoor views of native grasses and wildflowers. The vegetation is that of a reconstructed prairie, similar to the diverse mix of forbs and grasses that covered the Midwest until a blacksmith named John Deere introduced a new steel plow in 1837. Deere's plow enabled early settlers to turn the prairie soil with ease. (Previous wood and cast-iron plows were nearly useless on the prairie because the moist black topsoil stuck to them and had to be scraped off every few feet.) Millions of acres of tallgrass prairie disappeared under John Deere's plows in the 1840s and '50s. Today the only prairie preserves in and around Moline are the ones Deere & Company has reconstructed on its corporate properties.

At the Harvester Works Product Development Center I talked with Tom Hitzhusen, senior division engineer for future combines. He met me in the lobby of the building and ushered me into his office, a room large enough for a desk and a conference table. Reflecting the architecture of the building as a whole, one wall of his office was made of glass. From his desk he could look out over a large, open area divided by head-high partitions into a warren of workstations. The dividers were arranged so that a visitor like me could not see what went on behind them. The view was rigorously bland and uninformative, betraying no trace of the personalities within except for a small sign taped to the back of a computer that said: "Commit random acts of kindness and senseless beauty."

Hitzhusen was in his mid-forties, of medium height and stature, and wore a grayish-blue striped suit. His dark hair was cut straight across his forehead in bangs. He looked a bit like Mr. Spock, only warmer and without the pointed ears. One of the first things he did was to reach over to the credenza behind his desk and pick up a framed photograph of a shiny green-and-yellow tractor. Although the tractor looked as if it had just rolled off the assembly line, it was actually a 1946 John Deere Model A.

"I spend my days thinking about the future," Hitzhusen said. "I spend my evenings immersed in the past, restoring old tractors. Right now I'm working on a nineteen-thirty-one John Deere 'D.' It had water-cooled injection. Magazines today rave about a new feature on all the latest, hottest, motorcycles, boats, and things like that. You know what that new feature is? Water-cooled injection. Sometimes I wonder, is there anything truly new? A lot of what we do nowadays is recycling and refining old ideas."

I took this as a response, however unwitting and indirect, to the opinion Mark had voiced about Deere's Maximizer combine—that the new machines contain the same basic ingredients as old-fashioned threshing machines. That may be true, Hitzhusen implied, but the difference between the execution of those basic ideas then and now reflects several intervening revolutions in the development of materials, manufacturing methods, and product design.

The building in which we now sat, Hitzhusen told me, was completed in 1980. Before that the research-and-development staff for combines was housed at John Deere's Harvester Works, right in the factory. But here the product planners were moved out of the din and distraction of the factory and into a separate, quieter setting. "About one hundred ninety people work in this building," he said. "At least half of them are engineers of some kind." (The engineering ranks included Hitzhusen, who had both a bachelor's diploma and a master's degree in agricultural engineering.) "The rest of the staff includes designers, clerical workers, field mechanics or test mechanics." Two-thirds of these employees, or about 126 people, he explained, spent all of their time on combines; the rest worked on planters and seeders.

"When most people think of innovation," Hitzhusen said, "they think of a guy with kind of a crackpot idea. Everyone tells the guy it won't work, and that's what fuels him. He has a burning fire in the belly. I call that inventiveness. There's much less of that now. The thing that drives innovation today is not a fire in the belly but the twin desires for profitability and for meeting the customers' needs. We still innovate today, in a setting like this, but it's more sophisticated and businesslike. The maturation of this process parallels the maturing of the food system."

When I was arranging my visit with Hitzhusen I told him that I was curious to learn about the process of invention and innovation in a corporate setting. He had obviously given the matter some thought before our meeting and seemed eager to discuss it.

"The individual inventor's specialty is what I call 'leap design'—dreaming up dramatically new concepts," he said. "What we do in an operation like this is more incremental. We build on concepts that have proven their value in the marketplace. We do it quickly, using the latest and best tools for gathering data, moving it, analyzing and processing it. All these tools—CAD systems, computer modeling, electronic monitoring, market research—help us take risk out of the process. They help us improve on existing ideas. What they do not do, though, is create brand-new ideas. They don't create the spark. In fact, you can almost get trapped in a mountain of data. I can gather data to ten decimal points of precision, but then what do we do with it? That's why we're constantly going out into the field, talking with customers and users of these machines. We hold focus groups, consult our dealers, and get our hands dirty in the field with customers."

"So where do the new ideas come from?" I asked.

"A lot of them come from individual inventors and small shops," he said. "A big company like Deere can then evaluate them. If they look promising, we might either buy the patents or, in some cases, acquire a small company. For example, the basic ideas for our Max-Emerge planter came from a farmer. We bought the patents and developed them into a marketable machine. As another example, I once worked on a power-till seeding system. The original idea for it came from a professor at the University of Kentucky. Incidentally, we often fund university studies. It's a way of sowing some seeds and searching for ideas. The University of Kentucky professor sold us his rights to the seeder, then he went back to teaching and inventing, which is what he does best. He had no business sense and could never have brought a product to market. But we have a very detailed understanding of the market, plus manufacturing and marketing capability and an unrivaled network of dealers. The agriculture industry is well-served by these kinds of partnerships."

I asked if there was any truth to the allegations, often heard wherever brand-x loyalists or competitors gathered, that Deere & Company sometimes ignored the patent rights of individual inventors and either stole inventors' ideas outright or designed around their patents.

"No," he replied. "We are very cautious, very careful about that. We get many letters every day from farmers and inventors mentioning ideas to us that they think are new. They don't know it, but we've already seen ninety-five percent of those ideas. We may already have a long history of testing and developing some of them. But the farmer or inventor thinks it's original to him. If we ever come out with something similar, he may feel, incorrectly, that we've stolen the idea. We go the second mile to be good citizens in this respect. If one of these cases goes to court, in the eyes of most juries we're the deep-pockets company, and we'll probably lose. Obviously we don't want it to come to that."

Besides, he added, patents now play a smaller role than they used to as the currency of innovation. "The world has changed drastically, but patent law is still basically the same as ever. Patents are still in effect for seventeen years. Nowadays that represents several product cycles. Markets come and go in that time, so it may not be worth it to patent every little feature we come up with. Often it's not worth the expense and the hassle. We still encourage our engineers to take out patents. When they get one, we buy them a cup of coffee and say, 'Good job,' but we don't make a big deal out of it.

"We work in teams here," he said, leaning forward for emphasis. "Everything we do is done on a teamwork basis. We may have one team working on, let's say, a new header, and another one working on a seeder or on some specific area within a combine. In a team situation it's often difficult to determine whose idea something is. Everybody contributes. Rewarding individuals for patents creates an internally competitive environment that can stand in the way of teamwork. Your independent inventor type is typically more of a rugged individualist who doesn't always function well in a team-based environment. Inventive types are more patent-driven."

In a corporate setting, Hitzhusen said, the drive for profits was far more important than the quest for patents. "Most farm-equipment companies today are run by people with accounting and business degrees, not people with technical or even farming backgrounds. The top managers are bottom-line-driven. Last quarter's report dictates a lot. There is a need to survive and grow as a business, and that may mean there are fewer resources available for radical new-design thinking.

"Still," he added, "we invest more than most other industries in research and development—roughly a million dollars a day, in good times and bad. We constantly have to invest to make sure our machines are the best in their class. It doesn't necessarily pay to be light-years ahead. The best place to be is just one major step ahead of your competition."

Wasn't it difficult to innovate, I asked, with something like a combine—a very complicated, expensive machine that sold in relatively modest numbers?

"The free fall in combine sales, and in the farm-equipment industry generally, has changed everything about how we operate," Hitzhusen answered. "The total market for combines in North America went from about forty thousand a year in the late nineteen-seventies and early nineteen-eighties to less than ten thousand last year. Could GM or IBM survive if their market shrank to one-quarter of its original size in ten years? I doubt it. We had to resize our business to be profitable at a smaller scale.

"In a way, we contributed to that free fall in sales." I was a little surprised to hear a corporate manager admit such a thing so candidly. He went on: "Today's largest combine has ten times the capacity of the machines of fifteen or twenty years ago. The total amount of harvestable acreage in the country has not changed much in that time. Larger machines on the same amount of land means fewer machines will be needed. It doesn't take a rocket scientist to see that, and we knew it was coming. But we also knew we had to respond to the needs and desires of our customers. They wanted bigger machines with a higher capacity, and they wanted more reliable machines that wouldn't break down. That's what we provided for them. We gained in market share, and are now the leader in the industry by a significant margin. But the overall market shrank, and we are constantly adjusting to that."

"How can you adjust?" I asked. "I don't suppose you'll start making smaller combines that wear out faster."

"No," Hitzhusen said, smiling. "We'll continue to respond to our customers' needs. But we have to learn to drive our break-even point down, even as the machines get more complex. To do that, we have to innovate internally. The process of innovation is not always reflected in the product itself. Innovation must also occur in the process of designing and building the product. The whole process of developing products, from design to marketing, has been integrated and streamlined. The joke about how things used to be done was that the product engineers would throw a skunk over the wall to manufacturing and say, 'Okay, build it,' then manufacturing would throw the skunk over the wall to marketing and say, 'Okay, sell it.' Now everybody gets into the act from the beginning. That has helped us to shorten the product-development cycle, which is the period from the germ of an idea to a marketable machine. If you define the length of the old product cycle as one hundred percent, now we're at sixty-six percent. This allows us to put new technology into farmers' hands more quickly. It has accelerated the pace of technology."

To achieve the new efficiencies of streamlining and integration, Hitzhusen told me, companies like John Deere have had to invest enormous amounts of time and capital not only in retooling their factories but also in retraining their workers and managers. Much of that training has had nothing to do with technology per se. "We've spent more over the last several years on training in *human* skills than we have on training in technical skills," he said. "We're teaching our people to work together. We train them in things like team building, and in how to deal with conflict in a group. When you put a bunch of engineers together they want to start laying down lines and building iron. That's their instinct. But we send them off, away from the factory, away from the computers, and say, 'Before you build iron you have to build a team.' By getting the internal bickering out of the way, you release enormous amounts of energy for constructive teamwork. It also builds on the concept of empowerment. People function better when they feel empowered. That means not having lots of managers giving top-down orders but, instead, having a more open, participatory workplace.

"If you ask about innovation over the last five years, much of it has been technical, sure," he elaborated. "But most of the innovation has been in changing the corporate culture. We couldn't build the machines we build now if we operated the way we did twenty-five years ago. Let me give an example. We just introduced a whole new line of tractors—the seven thousand series. They embody a whole slew of new technical features that advance the state of the art by miles. But those tractors don't cost significantly more than the old models they replaced. That is a difficult thing to pull off, believe me. But there is more innovation embodied in those machines than meets the eye."

. . .

Mark and I drove back to Haven from the Custom Harvesters' Convention by way of Buhler, a dot on the map where there was a huge combine junkyard called Mike's Equipment. Mark loves a good junkyard the way some people love a good secondhand bookstore. Junkyards, for Mark, are rich sources of ideas and inspiration. They're bottomless mines of iron that come preassembled in all sorts of useful configurations, like Polhem's mechanical alphabet. Most of all, junkyards teem with memories and stories. Mike's Equipment contains several hundred combines—maybe a thousand—and every one of them has a story behind it.

"There's a nineteen-eighty Gleaner N," Mark said, pointing to one burned-out hulk. "The N series was real notorious for catching on fire. See how the engine is all enclosed in sheet metal back there? Chaff and dust would collect around the engine and exhaust manifold. The guy who owned that machine was probably combining wheat on a hot July day and all of a sudden he heard a noise and looked around behind him. Oh, shit! The whole back end of the combine was in flames. He got the hell out of there, I'll tell you that."

As we walked up and down rows of discarded machines, Mark singled out a rusting red combine, an International Harvester 503, and looked it over fondly. "This is a vintage nineteen-sixty-two or 'sixty-three International," he said. "A little puny thing like this don't look like much now, but they were the hogs back in them days when they first come out. Let's see if I can find a component on this old five-oh-three that's still on their latest model, the sixteen-eighty." He studied the machine closely, opening a side panel to expose its innards. All I could discern was a chaos of belts, pulleys, and rocker arms. But Mark saw the utterly familiar features of an old acquaintance. "Here's one," he said shortly. "The bushing on the sieve."

We moved on, slowly, quietly, as though we were walking through a cemetery and reminiscing about the dead.

"Here's a McCormick one-fifty-one," Mark said. "Man, that's an oldie. And look at that old Baldwin Gleaner over there. You could restore that one and put it in a museum."

We came to a newer machine—a John Deere 7700, roughly twenty years from the cradle and now laid to rest here in Mike's combine graveyard. "I remember when those first came out in the early seventies. I would have given my left nut to have one."

Suddenly Mark became engrossed in examining a large cast-iron assembly that lay on the ground. He explained that it was the bell housing from the rear of the engine on an International Harvester 915—the same model he and Ralph had bought from this junkyard a few days earlier. Mark flipped the clutch arm, spun the drive shaft, then turned the assembly on its side to look underneath it.

"I'll have to keep this configuration in mind," he said after a few moments. "We may use it someday. It could save us *beaucoup* bucks."

HEREDITY AND HORSEPOWER

Whoever could make two ears of corn or two blades of grass to grow upon a spot of ground where only one grew before, would deserve better of mankind, and do more essential service to his country than the whole race of politicians put together.

—Jonathan Swift, *Gulliver's Travels* (1726)

Oh, what a beauty she was. There was nothing more beautiful in the whole world than a brand-new tractor. . . .
"Got to keep up with the latest tricks," laughed Uncle Henry, "if you want to make a corn farm pay."

—Lois Lenski, *Corn-Farm Boy* (1954)

On a miserable December day in 1922, three men met in a cornfield a few miles north of Des Moines, Iowa. Everything—ground, trees, cornstalks—shimmered and crackled under a glaze of ice. Pellets of sleet fell on top of the ice, like BBs on a frosted mirror. Walking was treacherous, but the three men seemed not to notice. They had come to this cornfield like gladiators to the arena, their minds fixed on battle. A small group of onlookers gathered nearby, stomping their feet and clapping their shoulders in the cold as they waited for the bout to begin.

Each of the three principals wore a fingerless leather glove on one hand. Fitted with straps, buckles, and metal-studded plates on the palms, the gloves looked like implements of medieval torture. Three teams of draft horses, each hitched to a stout wagon, stood by, their breath steaming in the cold. The gladiators took positions by their chariots and waited. Someone gave a shout, and the three men lunged forward, disappearing into the cornstalks and the sleet. A chorus of dull thuds rose from the field, accompanied by the sound of the wagons slowly advancing.

The men were not killing one another; they were picking corn. As they moved down the rows, each combatant grabbed an ear, ripped it with his gloved hand from its husks and stalk, and threw the husked ears against the backstop on his wagon—all in one fluid motion. Millions of farmers picked corn by hand every autumn, but these three—Ben Grimmius, Louis Curley, and John Pederson—were extremely good at it. They had won the first round of a contest sponsored by *Wallaces' Farmer,* the well-known midwestern farm weekly, and they had come together on this December day for the championship round. Whoever picked the biggest load of cleanly husked corn in one hour would be the winner.

The cornfield, near a little town called Alleman, was in bad shape. At least half of the stalks in it had been toppled by ice and wind. Grimmius, Curley, and Pederson spent much of the hour bent double, retrieving ears from the ground. They struggled to keep their footing on the ice and squinted to keep the sleet out of their eyes. The few spectators were even gladder than the contestants when the timekeeper shouted that the hour was up. The loads were weighed, unshucked ears counted, and a winner declared: It was Lou Curley, who had gathered 1,100 pounds of corn in his wagon. Alleman seemed more like Mudville than Cooperstown that day, but it had just become the cradle of a new American sport.

The next year's contest, held a little earlier in the season, attracted more spectators and wider publicity. The year after that, contestants came from all over the corn belt. By the late 1920s, eleven midwestern states were holding their own state husking championships each fall. The winners advanced to the national finals, which participating states took turns hosting.

The National Corn-Husking Championship became one of the country's great annual festivals. At its peak in the 1930s, the so-called Battle of the Bangboards drew crowds of 160,000 spectators—more people than had ever attended a single sporting event. NBC Radio broadcast the contests live to millions of rapt listeners. Newsreel crews came in force. *Life* and *Newsweek* dispatched reporters and photographers, and so did dozens of local newspapers from all over the country. Blimps hovered, biplanes buzzed. Ford and General Motors displayed their latest automobile models inside big tents. Cornhusking was the main event, but there were also plowing demonstrations, tractor pulls, minstrel shows, corn-judging shows, even hog- and husband-calling contests. At the 1933 championship, held in Nebraska, veteran auto racer Barney Oldfield amazed the crowd by driving a farm tractor at sixty-five miles an hour, setting the all-time tractor speed record. President Franklin D. Roosevelt gave the starting signal for the 1936 contest from the White House via telephone. Champion huskers like Fred Stanek, Ted Balko, and Elmer Carlson were as famous and revered in rural America as Babe Ruth and Jack Dempsey.

Elmer Carlson, winner of the 1934 National Corn Husking Championship, picking his way to victory. (Wallaces Farmer)

The wonder is not that husking contests were so popular but that corn was still harvested mainly by hand until well into the 1940s. Corn was the most economically important crop in the United States, and had been since the turn of the century. "The value of this crop almost surpasses belief," wrote Secretary of Agriculture "Tama" Jim Wilson in the U.S. Department of Agriculture's 1908 *Yearbook.* "It is $1,615,000,000. This wealth that has grown out of the soil in four months of rain and sunshine, and some drought, too, is enough to cancel the interest-bearing debt of the United States and to pay for the Panama Canal and fifty battle ships." Even so, corn had missed out on agriculture's industrial revolution. Machines like the reaper, thresher, binder, and combine were used only for small grains like wheat, oats, and barley. Methods of growing and harvesting corn remained remarkably unchanged from the time of Columbus to the time of the first husking championships in the early 1920s. Then, rather suddenly, two inventions put not only corn culture but all of agriculture on a new footing. Those two inventions were hybrid corn and the all-purpose tractor. Without them, a combine of the size and capacity of Mark Underwood's would have little reason to exist.

CORN, KNOWN OUTSIDE of the United States as maize, originated in a pre-Columbian corn belt that extended from what is now New Mexico and Arizona across the Mexican plateau and deep into Andean valleys. Countless generations of indigenous peoples—Maya, Incas, Aztecs, Hopi, Zuni, Cherokee, and many others—planted, cultivated, and worshipped corn as their primary source of sustenance. Collectively, these ancient nations accomplished the most remarkable plant-breeding feat of all time: They transformed a single variety of teosinte—a wild Mexican grass bearing a few seeds in a small, cobless husk—into hundreds of domesticated varieties, none of which bear even a faint resemblance to their common ancestor.

No cultivated plant has assumed a greater variety of shapes, sizes, and colors, or has adapted to a wider range of growing conditions. Varieties of maize range in mature height from less than two feet to more than twelve; cob length varies from one inch to eighteen inches; kernel colors run from lily-white to yellow to scarlet to a blue as dark as a moonlit night. Some varieties can grow on less than five inches of rain a year; others thrive in places where the annual rainfall exceeds 170 inches. Maturity can arrive as soon as two months after germination, or it can take as long as two years. Corn will grow both at sea level and on alpine terraces at 11,000 feet and higher. One thing none of these myriad varieties can do, however, is survive and propagate in the wild. By the time Columbus set foot in the New World, corn was already as far removed from its wild ancestors in appearance and habit as a toy poodle is from a timber wolf.

European farmers of the American frontier were at first fairly undiscriminating about the kinds of corn they grew. They planted whatever locally adapted strains of Indian corn they could get their hands on. In the northeastern United States, this was usually a variety with short, slender stalks and white cobs that bore eight rows of smoothly oblate, hard kernels: flint corn. In the South, from Maryland and Virginia down to Florida, native tribes grew a different type of corn. This variety had tall stalks and big ears with sixteen to thirty rows of narrow kernels, pointy and soft. Settlers called this kind of corn gourdseed because the kernels resembled the seed of a dipper gourd. Gourdseed-type corns didn't taste as good (roasted, boiled, or ground) as the northern flints, but they yielded more abundantly in places where summers were long and wet.

White settlers took Indian corn in its cornucopian variety and began to remake it in their own image. The method was simple: A farmer walked out into a field of mature corn, chose the plants or ears that most struck his fancy, and set the prime ears aside in a dry place. These ears would provide seed for the next growing season.

What were the most desirable characteristics to look for when selecting seed ears? Every grower had his or her own idea. Some liked tall plants with many ears; others preferred medium-size plants that bore one or two shapely ears. Smooth kernels suited some growers; others favored a rough or dented texture. The ideal shape of the ear was a matter of heated debate. Should the ear be long and tapering, or stocky and square-shouldered? There was more unanimity on the question of which kernel color was best. White farmers (most of them) harbored an implicit, unexamined preference for pearly or pale-yellow kernels. They bred out any red, blue, purple, or calico tendencies whenever possible.

None of these aesthetic preferences had much, if anything, to do with the productivity of the corn. Taste and milling characteristics weren't given much consideration either, since most of the white settlers' corn was fed to animals. What mattered most was appearance. A consensus gradually developed about what constituted beauty in corn: slender, tapering ears of good size and heft, one or two per plant, each ear having about twenty straight rows of creamy yellow, smoothly dimpled kernels. The explosion of diversity that had occurred during the course of 10,000 years of indigenous agriculture gradually slowed, then stopped, then began to reverse in a genetic implosion whose logical conclusion was a kind of Aryan uniformity.

The unassuming architect of one master race of corn was a mild-mannered farmer named Robert Reid. In search of better farmland, Reid moved from southern Ohio to central Illinois in the 1840s. Among the possessions he brought with him to Illinois was a bag or two of gourdseed-type corn. Reid called the variety Gordon Hopkins, after the Virginia farmer who had given it to him. Although his Gordon Hopkins corn had done well in Ohio, it germinated poorly in Illinois. The field where Reid planted it was full of empty spaces where the kernels had failed to sprout. Reid replanted the empty hills with a short-season, eight-row flint corn that someone had brought from New England—a strain known as Little Yellow. Naturally enough, the two varieties fertilized each other. Pollen from the tall, shaggy Gordon Hopkins plants sifted down onto the silk of the Little Yellow ear shoots.

The offspring of this union—a varietal cross between a northern flint and a southern gourdseed—pleased Robert Reid. He and his son, James, planted seed from these ears the next spring, and refined the cross over the course of several seasons. When they selected seed, they chose early-maturing, plump, cylindrical ears, about ten inches long, with eighteen to twenty-four rows of perfectly straight, deeply dented kernels per ear. They favored a uniform yellow color and diligently bred out a reddish tendency that came from the Gordon Hopkins.

The Reids were aesthetes above all; they did not breed directly for high yield. As it happened, though, the strain of corn they produced not only appealed to them visually but was also highly productive. James Reid, carrying on the breeding work after his father's death, entered some of his finest ears in the Chicago World's Fair of 1893. His entry won the blue ribbon. "Reid's Yellow Dent," also known as "World's Fair Corn," spread rapidly across the corn belt in a blaze of fame, replacing hundreds, maybe thousands of quirky local lines with its vigorous though pallid uniformity.

Spreading Reid's Yellow Dent far and wide was the mission of Perry Greeley Holden. A birdlike man with a narrow face, round spectacles, and Ross Perot ears, Holden was a botanist with an impressive academic pedigree. At Michigan Agricultural College he had apprenticed under William James Beal, who in turn had studied with Asa Gray, the Harvard botanist who was Charles Darwin's closest American friend and colleague. Holden left Michigan, his native state, in 1895 to take a job as a professional corn breeder in Illinois. There he encountered samples of the corn that had won the Chicago World's Fair prize a few years earlier. Holden included some World's Fair Corn in his field trials and immediately concluded that it was superior to any corn he had seen before. It yielded well, but it also possessed a certain je ne sais quoi that made a corn connoisseur such as himself tremble with delight. Like most other corn experts of his day, Holden believed that corn had an almost ineffable visual quality that signified its productive potential—an aura of fecundity, like the "dairy glow" of a vibrant milk cow.

In 1902 Holden went to Iowa to join the faculty of Iowa State College. He was dismayed to see that Iowa farmers were planting a motley assortment of corns, including some very sorry-looking ones. Holden vowed to change that. When he wasn't teaching at Iowa State, he traveled from farm to farm in a wagon proclaiming a message of better living and better farming through better corn. People started calling him the corn man, or the corn evangelist. He taught farmers how to test their seed for germination to make sure it would sprout. He taught them how to select only the finest ears for seed. And he passed out samples of Reid's Yellow Dent.

Holden approached officials of some railroads that served Iowa, asking for help in his quest to educate Iowa farmers about corn. According to some accounts, the railroad officials brushed Holden aside until "Uncle" Henry Wallace, the influential editor of *Wallaces' Farmer,* intervened on Holden's behalf. Wallace persuaded the officials that a more plentiful corn crop would benefit the railroads: More corn meant that more hogs and cattle—paying customers—would be fattened and shipped to market. A great many of these hogs and cattle, and some of the corn, would be transported on the railroads to the great stockyards and elevators of Chicago.

The Chicago, Rock Island & Pacific Railroad was the first to embrace this logic. In 1904 it furnished Holden with the use of a five-car train. Officially, this train and its successors were called the Iowa Seed Corn Specials. In the Iowa hinterlands, they came to be known as the Corn Gospel Trains.

Holden was the chief evangelist on the Corn Gospel Trains, but he had disciples. Most of them were agronomists from Iowa State College. They rode in two private cars at the front of the train, reserving the final three cars for lecture halls. In the spring of 1904, the first Gospel Train covered the entire state of Iowa, stopping at 570 towns. At each stop scores and sometimes hundreds of farmers waited to board the train and hear "the sermon." The sermon took about thirty minutes and was preached simultaneously in each of the three "classroom" cars.

The essence of the corn gospel (as Professor Holden laid it down) was that Iowa's corn crops, on average, fell sadly short of their potential, and that the reason for poor crops was poor seed. But salvation was at hand: Farmers could improve the quality of their seed corn by testing it for germination and by learning what characteristics to look for when selecting corn for seed. Prizewinning ears of Reid's Yellow Dent were held up like sacred objects—examples of the ideal that farmers should strive for.

Farmers hear "the sermon" on a corn gospel train, 1905. (Iowa State Univ. Library/University Archives)

Holden and his disciples preached their sermon as often as twenty times a day. After three months, more than 110,000 farmers had heard it. Railroad companies began to regard Corn Gospel Trains as tools for improving not only their business but also their public relations. (Many farmers viewed railroad owners as greedy monopolists who bled the profits from agriculture.) The railroads dispatched more corn trains in following years, and the number of farmers who heard the gospel multiplied.

Another vehicle Holden used for spreading his message was the corn show. For the first four decades of the twentieth century, corn shows were common and popular events all over the Midwest. Held at every county fair, state fair, and local harvest festival, corn shows were essentially beauty contests. Farmers would enter ten of their best-looking ears of corn in the contest, displaying them in wooden racks, side by side. A panel of judges would examine the entries and rate them on scorecards. Points were awarded mainly for uniformity—uniformity of ear size, kernel size, color, shape, and texture. The highest-scoring entries won ribbons and trophies. Corn shows became a major preoccupation in thousands of communities. They served, in part, as a formalized outlet for competition among farmers, and also as a source of neighborly diversion.

Holden wrote the book on corn shows—a manual in which he spelled out the rules for entry and the standards for judging. He justified the whole enterprise in the name of corn improvement, taking for granted a direct relationship between cosmetic perfection and economic value. "The best ear of corn for seed," he wrote, "is also the best ear for the show."

HENRY AGARD WALLACE, "Uncle" Henry's grandson, was a teenager when Perry Holden started barnstorming across Iowa in the Corn Gospel Trains. Young Henry (which he was called all his life, to distinguish him from his grandfather Henry and from his father, Henry Cantwell Wallace) was an exceedingly earnest young man with a long, narrow face, penetrating eyes and a swept-back forelock that no amount of hair oil seemed able to tame. He loved botany, an interest he acquired early in life from his mother, who was an avid gardener, and from George Washington Carver, who took young Henry under his wing in the early 1890s when Carver was a student at Iowa State

College and Henry's father was on the college faculty. Young Henry took all his interests quite seriously, projecting an aura of middle-aged sobriety from the age of about six.

As a teenager, Wallace held mixed feelings about the Corn Gospel Trains. On the one hand, he admired Perry Holden's dedication to helping Iowa farmers improve their principal crop. On the other hand, young Henry had doubts about some of the professor's teachings. In particular, he questioned Holden's assumption that appearance was a reliable indicator of productivity in corn. As a friend of the Wallace family, Holden occasionally paid visits to Uncle Henry's house, so young Henry had opportunities to discuss his ideas with the distinguished teacher. The young man expressed doubts about the practical value of corn shows, and wondered if judging by scorecard standards might even have a negative effect by encouraging farmers to breed for cosmetic traits—traits that might have nothing to do with economic value. Would Professor Holden's pretty show ears truly outyield corn of lesser beauty? Wallace was polite but persistent with his questions. Finally Holden said, "If you don't believe me, why don't you check my judgments?" He suggested that young Henry perform a scientific yield test.

To Henry Wallace, who was sixteen at the time, no challenge could have been more appealing. Holden supplied the boy with thirty-three ears that had been entered in a corn show. All the ears looked very much alike, but Holden, with his practiced eye, had ranked them from best to worst. Henry carefully labeled the ears according to Holden's ranking, shelled them, and planted the seed in a five-acre plot.

Five acres of corn is a lot for one person to plant, thin, cultivate, hand-pollinate (to prevent unwanted cross-fertilization), and then harvest—all by hand. Henry Wallace did it methodically, conscientiously, even lovingly. After harvesting his plot, he weighed the test ears and extrapolated a yield-per-acre for each one. The yield figures varied from thirty-three to seventy-nine bushels per acre—a surprisingly wide range, considering how similar the original ears had looked. The ear to which Holden had awarded a blue ribbon ranked near the bottom in Wallace's yield test; conversely, the winner of the yield test ranked as one of the least perfect ears by Holden's standards. In general, Wallace's test showed that there was no relationship between corn-show perfection and yield. Describing his test in *Wallaces' Farmer* a few years later, young Henry made the rather shocking observation that homely-looking nubbins might make better seed corn than the fine, sleek ears that won ribbons at shows. After all, he wrote, "what's looks to a hog?"

HENRY WALLACE PERFORMED his first yield test during a period of revolutionary discovery in the new science of genetics. He turned twelve in 1900, the year Gregor Mendel's work was rediscovered after it had languished in obscurity for a third of a century. Mendel, who lived in an Augustinian monastery in Brünn, Moravia (now part of Czechoslovakia), began a series of experiments with garden peas in 1856. He observed the inheritance of biological traits with each generation of peas,

The young Henry A. Wallace shows a handful of cross-fertilized corn kernels to his grandfather, "Uncle" Henry, 1913. (State Historical Society of Iowa)

making a careful tally of how often traits like tallness, dwarfishness, and red or white blossom color turned up after various controlled matings. Interpreting his data, he developed some novel theories about the nature and mechanism of inheritance. He presented these ideas to the Brünn Natural Science Society in 1865 in a paper titled "Experiments in Plant Hybridization."

The gist of this paper was that individual traits, such as red or white blossom color in garden peas, do not blend like cans of red and white paint. Rather, Mendel claimed, these traits are transmitted by paired elementary units ("elements," he called them; the word *gene* came later), which separate and independently recombine during sexual reproduction. Furthermore, the possible patterns of recombination occur with a frequency that can be predicted mathematically.

In Mendel's day, mathematics was like a foreign language to most investigators of the living world. Natural science consisted almost entirely of description and classification. So when Mendel read his paper to the members of the Brünn Natural Science Society, nobody understood it. But Brother Mendel was a respected member of the society and his paper was duly published in the group's proceedings. There it remained, unread—or at least unappreciated—until sixteen years after Mendel's death. Then, in 1900, three European botanists independently produced experimental results and theories that were, it turned out, basically the same as Mendel's. The three scientists, in searching the literature, were astonished to find that a Moravian monk had anticipated their discoveries by a generation.

By the time Mendel's work resurfaced, biologists were starting to appreciate the value of numbers. The new generation of life scientists was far more receptive to, and conversant in, quantitative ideas and methods than the contemporaries of Mendel and Darwin had been.

In that regard, Henry Wallace was a modern biologist from a young age. George Washington Carver taught him, from the age of six or seven, how to identify different species of grasses; at the same time, young Henry was developing a keen interest in mathematics. The meticulous way in which he carried out the corn-yield test on Perry Holden's show ears demonstrated his knack for applying statistical methods to the study of plant (and, later, animal) populations. When he entered Iowa State College in 1906, he was already thinking in ways that some of his older professors found excessively modern.

While he was an undergraduate, Wallace began to think about plant and animal breeding in terms of the new Mendelian framework. He knew that inherited traits, as Mendel had said, were determined by the combination and recombination of discrete genetic units. Some physical features, like the blossom color of the garden pea, were simple traits, controlled by a single genetic unit. Other characteristics were extremely complex, controlled by many different genetic units acting in symphony. The yield of a corn plant, Wallace understood, was one such complex characteristic. Mendel's algebra of inheritance was clean, elegant, and precise when applied to single-gene traits. Complex, multigenic traits, however, were another matter.

Highly evolved organisms, such as corn, were giant grab bags of genetic possibility. You might cross two high-yielding strains of corn, but the genes governing yield occurred in combination with such a hodgepodge of other things that the outcome of the cross was unpredictable, a huge toss-up. Wallace found the pure science of genetics fascinating, but he wondered if it could ever be of practical use in the messy, multigenic world of corn breeding.

A THOUSAND MILES east of Ames, where Henry Wallace had started his freshman year at Iowa State, a geneticist named George Harrison Shull began a series of experiments that would bring Mendel one large step closer to the farm. Shull was an Ohio farm boy turned botanist. Shortly after he received his doctorate from the University of Chicago, one of his former teachers, Charles B. Davenport, invited Shull to take a position as a staff scientist at the Carnegie Institution's new Station for Experimental Evolution at Cold Spring Harbor, on the north shore of Long Island.

Both Davenport and Shull were strong proponents of Mendelism. At Cold Spring Harbor, they hoped to educate the public about basic genetic principles while also advancing the science with sophisticated research. Davenport asked Shull to plant an exhibition plot that would demonstrate the law of segregation—Mendel's principle that genetic traits don't mix but are transmitted as separate and distinct units according to the rules of inheritance. Shull decided to demonstrate the principle by using starchy and sugary kernels on ears of corn, since it was easy to tell the two types of kernels apart just by looking at them.

Shull had no particular interest in corn as such (perhaps because he had seen enough of it as a farm boy in Ohio). He did, however, find corn an apt subject for genetic investigation. Many other plant species have tiny flowers with inaccessible male and female parts, making it difficult for humans to intervene and control fertilization. Corn, on the other hand, has separate male and female inflorescences: the male tassels, which shed pollen, and the female ears, which receive the pollen and nurture the kernels while they develop into seeds. Both the male and the female parts on corn are large, accessible and therefore easy to manipulate.

On its own, corn tends to cross-pollinate. Wind blows grains of pollen from the tassel of one plant onto the ear silks of other plants downwind. Seed that is produced in such a windborne sexual free-for-all is said to be "open pollinated": No attempt has been made to cross one specific plant, or group of plants, with another. But the physiology of corn makes it fairly easy to curtail the plant's natural promiscuity and enforce a policy of arranged matings. You do this by using paper bags, which serve, in effect, as iron girdles for the feminine silks and condoms for the masculine tassels. Just before pollination, you slip bags over tassels and ear shoots, sealing them at the bottom with staples. When the tassels shed, the bags hold the pollen. To cross plants A and B, you take some pollen from plant A's tassel bag and shake it onto the silks emerging from the ear shoots of plant B. (Or, if you wanted to, you could cross plant A with itself, thereby creating an inbred.) Then you slide a paper bag back over the fertilized ear shoot to keep out unwanted pollen.

The process went quite smoothly for George Shull the first year he tended his demonstration plot. He had found strains of corn that grew fairly well in the maritime climate of Long Island. The demonstration went so well that he decided to use corn as the test subject for his own genetic experiment.

Shull wanted to study the inheritance of "quantitative characters," or multigenic traits. The trait he chose to study was the number of rows of kernels on each ear of corn. This number naturally varies from fewer than ten to more than twenty. Shull wanted to plant strains that had varying numbers of rows, cross them in different combinations, and observe how each cross affected row count. Among the mating patterns he wanted to try was inbreeding—that is, crossing some plants with themselves.

The idea of inbreeding had always been associated with degeneracy and depravity. Any husbandman knew that matings between closely related chickens, pigs, or cows tended to produce small, weak, and sometimes deformed offspring. During the 1870s, Charles Darwin had tested the effects of inbreeding on plants. In a greenhouse at his home south of London, he crossed dozens of plant species and varieties. When he self-pollinated plants that were naturally cross-pollinated, such as corn, Darwin observed that the offspring were, for the most part, sickly and fragile. But when he crossed two plants of widely diverse lineage, he noticed that the progeny were strikingly large and healthy. Darwin thus became one of the first investigators to observe and describe the phenomenon of hybrid vigor.

Darwin's work lent scientific weight to the taboo against inbreeding. After the turn of the century, however, a few investigators, including Shull, began to look at Darwin's findings on inbreeding in the new light of Mendelism. While it was true that inbred offspring were feeble, perhaps inbreeding might have its Mendelian uses. Maybe it could be used as a tool to produce "pure lines" that would breed true for complex traits. In other words, inbreeding might be used to isolate certain combinations of genes in pure form, so a complex trait would be transmitted from one generation to the next with Mendelian predictability, as if it were a simple trait. This was one of the propositions that Shull set out to test with his corn patch at Cold Spring Harbor.

The laboratory where Shull and Davenport worked was in a new building made of brick and stucco, in the style of the Italian Renaissance. The two-story building stood near the edge of the harbor at the base of a wooded hillside. The setting was beautiful, but there was little room between salt marsh and hill for experimental plantings. Making the best of it, Shull converted the lawn around the lab into a cornfield. The domestic grass, *Zea mays,* grew side by side with wild saltmarsh grasses. Swans, Canada geese, and mallards swam only a few yards away as Shull tended his corn.

Shull's investigation began in 1905. During the next few years his inbred lines grew weaker and weaker to the point where propagation became difficult. He gathered data that supported his "pure line" theory concerning multigenic traits—but observed something else that was potentially even more interesting. When he crossed two genetically dissimilar inbreds, he found that the feeble parents produced offspring of amazing size and vitality. These hybrids yielded something like 20 percent more grain than average, open-pollinated plants. He was interested mainly in the expression of the row-count feature in these crosses, but he recognized that the great vigor of the hybrids had practical implications for agriculture.

He discussed some of these implications in a paper he read to members of the American Breeders' Association at a meeting in 1908. An ordinary field of open-pollinated corn, he said, was really a huge collection of uncontrolled and impossibly complex hybrids. Breeding open-pollinated corn by selection (which is what corn breeders did in those days) was therefore an extremely imprecise way to accomplish a particular result, because you were mating one complexly varied parentage with another. The desired outcome was by no means assured. Better to reduce this massive game of genetic roulette to a more manageable, predictable set of outcomes. This could be done, Shull suggested, by inbreeding corn strains, thereby isolating pure lines that, unlike open-pollinated varieties, consistently bred true. If you then crossed two feeble inbreds, you would get offspring that not only had predictable genetic features but also possessed fantastic hybrid vigor.

With that, George Shull laid the foundation of modern agricultural genetics. When he died in 1954, his obituary credited him with the invention of hybrid corn. *Time* magazine called him "the Santa Claus of hybrid corn." He did look remarkably like a Burl Ives version of Santa, but the implication that he single-handedly dropped hybrid corn into the world's Christmas stocking was exaggerated. Just as Charles Darwin had his Alfred Russel Wallace and Isaac Newton had his Gottfried Leibniz (contemporaries, that is, who came up with similar ideas at the same time), Shull had his Edward Murray East. Working independently in Connecticut, East covered much of the same experimental ground Shull did and arrived at many of the same conclusions. Shull foresaw the practical value of using inbreds to create hybrids before East did (although East was interested in corn as an agricultural crop, not just as a kind of botanical lab rat). A debate, sometimes heated, would develop over who made the more important contribution, Shull or East. The debate still simmers today. East, at any rate, credited Shull with giving him an important insight, which then became the basis for practical corn genetics.

When Shull first enunciated his ideas in 1908, however, an apparently unbridgeable chasm still lay between his theories and the widespread use of hybrid seed corn in agriculture. The problem lay with the inbreds. It took at least three generations of self-fertilization to create an inbred that would breed true. But third-generation-inbred seed produced terribly weak, unproductive plants. If the plants could be coaxed to maturity, they bore runty, misshapen ears, sparsely filled with kernels. Using such pathetic plants to produce commercial quantities of hybrid seed, and at a reasonable cost, seemed impossible.

Critics of Shull's ideas, meanwhile, popped up like spring weeds. For example, G. N. Collins, a respected researcher for the U.S. Department of Agriculture, warned against the perils of inbreeding, saying that it was "particularly dangerous" when applied to corn. The backlash against Shull was probably, in large part, a reaction of the corn establishment—USDA and agricultural-college types—against an outsider, a pure scientist whose interest in corn was merely academic.

Shull felt called upon to answer his critics, so he and his wife traveled to Omaha a few weeks before Christmas in 1909 to attend the National Corn Exposition, the corn establishment's big winter meeting. The Shulls spent their first day in Omaha setting up an exhibit of corn ears that Shull had picked from his experimental plot at Cold Spring Harbor only a few weeks earlier. They set out a few pairs of parent inbreds—small, shriveled, and pathetic-looking. Next to these parents, the Shulls displayed their hybrid offspring—great big, strapping ears. No demonstration of the kind had ever been seen in public before. The Shulls' small exhibit was swallowed up by baroque corn palaces and festive concessions at the Omaha exposition hall. Even so, the Shulls'

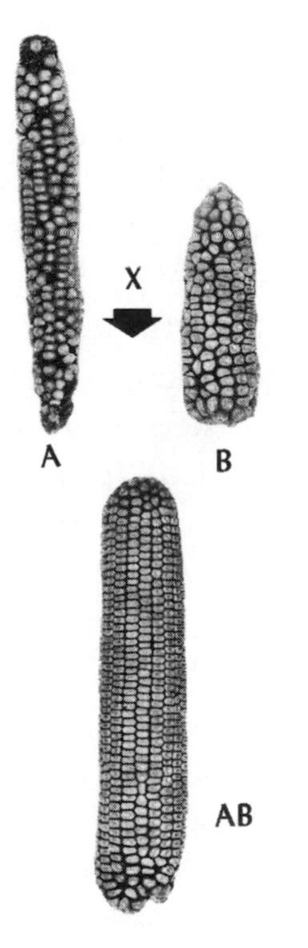

These two runty, gnarled inbreds, when cross-pollinated, produced the hybrid ear at the bottom.
(reprinted from Corn and Its Early Fathers *by H. A. Wallace and Wm. Brown, ISU Press, 1988)*

little display attracted large crowds of spectators who were astonished at this graphic demonstration of hybrid vigor.

At a general meeting during the exposition, Shull delivered a strong defense of his "pure-line method" and the practical benefits that it might bring to corn breeding. The speech marked a turning point in the careers of some of those who heard it. One scientist, Dr. N. E. Hansen of South Dakota, rushed to the dais after Shull finished his talk. Hansen slapped Shull on the back and pumped his hand. "You have all the other corn breeders skinned a mile," he said.

"I heard Shull talk in Omaha," recalled another scientist, T. A. Kiesselbach of the University of Nebraska. "I was impressed, amazed, challenged, enthused. It was an ear-opener. . . . Shull told of his experiment in which the poorest hybrid outyielded the best open-pollinated variety by ten bushels an acre." Kiesselbach and his colleagues from other agricultural institutions went back and started inbreeding programs at their home institutions. Breeders who previously thought that crossing inbreds was a harebrained notion were forced to admit that it might not be so far-fetched after all.

HENRY WALLACE READ Shull's Omaha address in the *American Breeders' Magazine* while he was still an undergraduate at Iowa State. The paper had the same effect on Wallace that it had on T. A. Kiesselbach. Young Henry wanted to test Shull's fascinating theories on his own, but he was finishing his senior year as an animal-science major and had no time for extracurricular plant breeding. In 1910 he was graduated at the top of his class, then he set out to see his home state on a long walking tour. He spent the next summer touring Europe. Late in 1912 he settled down in Des Moines to work for his father and grandfather at *Wallaces' Farmer*, the family paper. Finally, in the spring of 1913, he was able to start an experimental corn plot at his parents' house on Cottage Grove Avenue in Des Moines. That summer he tried self-pollinating, or "selfing," corn plants for the first time.

His trials had a practical goal: Someday he hoped to start a small seed-corn business that would supplement his modest salary as a lowly editor. He had just begun courting Ilo Browne, the daughter of a well-to-do farmer who lived near the town of Indianola. Wallace knew he would need more money than he could earn at the newspaper if he wanted to marry her and start a family. His dream was to produce a line of corn that would consistently beat Reid's Yellow Dent, not in appearance—he didn't give a corn peg about that—but in yield. Because of the seeds' yield advantage, he could sell it at a premium price, thereby earning some money for himself and helping farmers at the same time.

Despite his interest in the work of Shull and East, Wallace thought inbreeding was an unlikely path to his goal. Inbreds, he was discovering firsthand, were extremely difficult and time-consuming plants to work with. They were as finicky as hothouse orchids and took years to develop. And there was no guarantee that any given inbred would prove useful. Even if he discovered an inbred cross that produced a high-yielding hybrid, there was the problem of producing enough seed with the feeble parents to make the whole effort worthwhile. He soon abandoned inbreds as a hopeless cause and went back to crossing open-pollinated varieties.

Wallace seemed happier in a field of corn than he did in a roomful of people. Like many midwesterners, he drew a deep pleasure from the mere sight of tall cornstalks and waving tassels. His feeling for the crop was intuitive, intimate. Henry's brother James recalled a typical moment around 1918: "I can well remember him in the early morning, standing in his bare feet out in the small corn in his backyard reading his morning paper. Just looking at the corn in between times and reading the paper in between times and observing how it was doing."

Young Henry's breeding program eventually outgrew suburban gardens, so he moved his test plots to a farm near Johnston Station, a village less than ten miles northwest of the Iowa capital. (Ilo Browne, whom Wallace married in 1914, had traded a farm she inherited for the property in Johnston.) There he could try hundreds of varietal crosses every year. He mated all the standard North American varieties with one another and also with foreign strains from Russia, Australia, Hungary, China, and South America. The results were mostly disappointing. "Oftentimes these varieties would cross to produce unusually good results," he later wrote, "but when I tried to repeat the result the following year I oftentimes had bad luck." Varietal crossing appeared to be a dead end for reasons Shull and East had described. The alternative was their proposed path of inbred crossing.

The Shull and East doctrines received a boost in 1918 when Donald Jones, one of East's disciples, developed a new method for producing hybrid seed. Jones's method, called the double cross, established the procedure later used to create commercial hybrids. The double cross required four inbreds—A, B, C, and D. These four were mated in pairs—A × B and C × D—that yielded two kinds of hybrid seed, though in small quantities. This seed was planted the following spring, this time in alternating rows—one row of seed from the A × B cross, then a row of seed from the C × D cross, then another row of A × B, and so on. When the plants were about to sprout tassels and silk in mid-July, all rows of one of the parents—the A × B cross, let's say—were detasseled. By this process of emasculation, A × B became the designated female. This left the C × D plants as the only pollen-shedders in the field—the designated males. The C × D tassels naturally pollinated the A × B ears. (The male parents also pollinated their own ears, though these "selfed" ears were of no use to the corn breeder, and usually wound up in the hog trough.) The ears that developed on the female plants represented the hybrid (A × B) × (C × D). These ears were harvested, shelled, cleaned, bagged, and planted or sold as hybrid seed.

Obviously this two-stage process was more laborious than a single cross, and it required twice as many inbreds. Even so, the huge increase in seed produced during the second cross, with its attendant burst of hybrid vigor, more than justified the extra time and effort.

With Jones's discovery, corn hybrids looked as if they might become commercially practical after all. Henry Wallace learned about the double-cross method and looked at inbreds with rekindled interest. He monitored inbreeding programs at land-grant colleges and agricultural experiment stations, obtaining samples of any inbreds that interested him. Then he crossed these inbreds in novel combinations. In one such cross he mated an inbred out of Leaming, a well-known open-pollinated variety, with an inbred from an Oriental variety called Bloody Butcher. Bloody Butcher was named for its color—a vivid, coppery red. The offspring of this cross had yellow kernels, and they in turn produced impressive quantities of red ears. Wallace called his new hybrid Copper Cross. He entered it and a few other hybrids he had developed in the Iowa Corn Yield Contest at Iowa State in 1922 and again the following year. Copper Cross placed well in both years. Then, in 1924, it took the gold medal, becoming the first hybrid ever to win the prestigious award.

George Kurtzweil, a salesman for the Iowa Seed Company, saw Wallace's hybrids exhibited in the old armory building in Ames, where the yield contest was held. Wallace was explaining the finer points of hybrid genetics to Kurtzweil when the salesman cut him off. "This is all very interesting, Henry," he said, "but what I want to know is, when can we begin selling the seed?"

"Which one of the hybrids you see here would you like to sell?" Wallace replied.

Kurtzweil said the red one. Because of its novel color, he reasoned, farmers could more easily compare it with their other corn.

Wallace had only enough parent stock left to plant about an acre. The yield from this single-crossed acre was a mere fifteen bushels of hybrid seed, testimony to the weakness of the parent inbreds. (A good yield of normal corn, by contrast, would have been sixty, eighty, or more bushels to the acre.) Wallace wrote a few paragraphs about Copper Cross that appeared in the Iowa Seed Company's spring catalog. It was the first advertisement for hybrid corn ever published. "The Iowa Seed Company is offering Copper Cross—a cross of two inbred strains—as a foretaste of what is coming in the corn breeding world," Wallace wrote. "Ten years from now hundreds of crosses of this sort will be on the market. If you try it this year you will be among the first to experiment with this new departure, which will eventually increase the corn production of the U.S. by millions of bushels."

To many readers these paragraphs probably sounded like science fiction. Few people had as much faith as Wallace did in the new kind of corn. But perhaps because it was such a novelty, the whole inventory of Copper Cross sold out quickly at $1 a pound. Some of the farmers who bought it later reported that Copper Cross was the best corn they had ever grown. Since it was a single cross, however, it was a poor candidate for commercial production. In fact, Copper Cross was never offered for sale again.

Nonetheless, the episode convinced Wallace that it was time for him to start his seed business. He persuaded seven friends and relatives (including Ilo, his wife; James, his brother; and Jay Newlin, the capable tenant on the Wallace farm near Johnston Station) to invest in his new venture. The business was chartered in 1926 as the Hi-Bred Corn Company. Wallace and the other owners later prefixed the word "Pioneer." This addition to the company's name seemed appropriate: Pioneer was the first company to stake a claim in the newly opened territory of hybrid corn. For years Pioneer struggled like a settler on a remote frontier, waiting for the rest of the world to catch up.

. . .

IN 1910, the year Henry Wallace first read about the wonders of inbred crossing, another revolution was brewing—one that had nothing to do with genetics, though it would, in time, complement corn hybrids in a perfect marriage of genes and machines, heredity and horsepower. The tractor, along with the combine and the hybrid seed, would eventually stand as defining symbols of twentieth-century agriculture. But tractors as we know them took time to develop.

For a number of years, starting in 1908, the premier demonstration of power farm equipment in the world was a series of plowing trials held each year in Winnipeg, Canada. By 1908 Winnipeg had become the gateway to a huge new empire of wheat that developed (at the expense of California's wheat trade) in Canada's prairie provinces soon after railroads made the northern plains accessible to world markets. Conditions for growing a fine, hard spring wheat in districts such as Manitoba's Red River valley were perfect. The only thing lacking was enough labor and

A typical steam traction engine. (U.S. Department of Agriculture)

horsepower to plow fields that often stretched for miles in every direction. A single field here might be larger than ten midwestern corn-and-hog farms put together. The owners of these fields favored big steam traction engines for tillage. The largest of the "steam plows" they used weighed something like twenty-five tons and could turn furrows across a thirty-foot-wide swath in one pass. A big steam plow took the place of forty teams of horses and required relatively little human labor, which was especially scarce on the far-northern prairies.

The annual Winnipeg plowing match was a showcase for the machines that made this kind of farming possible. Manufacturers from all over North America entered their biggest and best traction engines—"tractors," as some people were beginning to call them. Tens of thousands of spectators from around the world came to see the big machines compete, and the results were telegraphed to every trade and agricultural center in the Western Hemisphere.

By the time of the first Winnipeg trials in 1908, steam-powered tractors from such venerable old companies as J. I. Case, Avery, and Aultman Taylor were beginning to face a challenge from the new breed of gasoline tractors. Internal combustion had been slower to mature than steam power. The German inventor N. A. Otto had developed the first practical internal-combustion engine in 1876. The Otto engine derived its power from a controlled explosion of fuel *inside* the cylinder rather than from the expansion of steam heated by a flame outside the boiler, as with a steam engine.

Internal-combustion engines made far more efficient use of fuel than steam engines did and were thus, at least theoretically, better sources of mechanical power. With a series of improvements made in the 1880s and '90s, the internal-combustion engine began to realize its promise. By 1899 there were about a hundred companies manufacturing gas-powered engines in the United States. Most of these engines were intended to sit in one place, providing belt power for washing machines, small corn shellers, and light industrial applications. An Iowa thresherman named John Froelich mounted a stationary gasoline engine on the frame of a steam tractor. In 1892 Froelich took this mechanical chimera on a fifty-two-day threshing circuit in South Dakota. R. B. Gray, the eminent tractor historian, wrote that Froelich's machine was "probably the first gasoline tractor of record that was an operating success."

Gas-powered tractors were still rare, exotic machines in 1908, when seven of them participated in the Winnipeg trials, along with a similar number of steam tractors. The two classes competed in separate categories, although nearly everyone saw the event as a showdown between steam and internal combustion.

It would have been difficult to resist comparing the two kinds of machines, and the comparison favored gas on most counts. Steam tractors were in general far larger and heavier than the gas-powered ones. The basic components of a steam engine—the boiler, firebox,

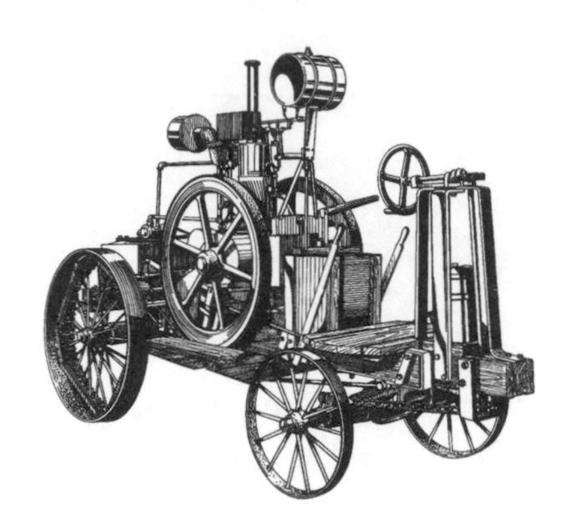

The first functional internal-combustion tractor, built by John Froehlich in 1892. (Deere & Co.)

flue, and pistons—had to be massive in order to generate a significant amount of power. An internal-combustion engine could generate the same amount of horsepower in a far smaller package. Gasoline, as a fuel, was also a good deal more compact and easier to handle than the coal and wood used in steam engines. One fill of the gas tank kept a gas tractor running all day, whereas a steam engine required a worker who did nothing but keep the firebox fueled and stoked. Steam engines also consumed large quantities of water, which had to be hauled constantly by an extra wagon, team, and driver. Finally, there were the dangers of boiler explosions and grass fires, which could be sparked by glowing embers that often shot from the smokestacks of steam engines. Internal-combustion engines were by comparison quite safe.

The advantages of gas tractors, however, were still largely theoretical in 1908. Steam traction engines had been around since before the Civil War. Despite their lumbering size and awkwardness, they were reliable performers. The same could not be said of gas tractors, which were notoriously fickle. Perhaps their most annoying characteristic was a stubborn refusal to start. It could take an hour or more of cranking and cursing to get the motors running. Once they were started, their operators often left them running all night to avoid starting them again the next morning. (Steam engines took a while to stoke up, too, but this was a routine and predictable delay.)

Partisans of steam made much of these foibles of internal combustion. An intense rivalry developed, and sometimes erupted into overt hostilities that for some old-timers were reminiscent of the reaper wars of the 1860s and '70s. Arguments about which was superior, steam or gas, occasionally deteriorated into fistfights. The president of one gas-tractor company claimed that steam-engine manufacturers refused to ship their machines in the same freight cars with his tractors.

Steam held its own during the Winnipeg trials of 1908 and 1909, not so much because of the superior strength of the steam entries as because of the failings of their rivals. By the time of the Winnipeg contest of 1910, however, the gap in reliability had narrowed. Improved systems of ignition, carburetion, cooling, and air-cleaning had all but eliminated the major bugaboos of gas tractors, making their advantages over steam traction engines increasingly evident. Also, with the discovery of vast quantities of oil in Texas, there seemed no end to the supply of cheap, convenient fuel for internal combustion.

If anyone doubted that the future of gas tractors was secure, the third Winnipeg competition settled the question. "The 1910 demonstration marked the first surrender of steam to gasoline," wrote Cyrus McCormick, the reaper inventor's grandson. In agriculture, as in industry and transportation, the age of coal gave way to a new age of oil.

AMONG THE THRONG of spectators at the 1910 Winnipeg Agricultural Motor Competition was Henry Ford. Always keenly interested in agriculture, Ford grew up on a farm in Michigan, where he developed a talent for tinkering with machines, as well as a distaste for farm work. One day during his thirteenth year, he and his father were riding in a buckboard wagon when they came upon a thresherman's steam engine creeping down the road under its own power. From then on, self-propelled vehicles fascinated Ford. His fascination became an obsession, and found its main outlet in the automotive field. But Ford never lost his interest in the machinery of farming. Some of his automotive innovations, he thought, might be transferable to the realm of agriculture. This idea no doubt lay behind his decision to attend the Winnipeg trials. He watched as twelve gas competitors and six steam-powered rivals turned wide swaths on Manitoba's immense, flat prairie. The weather had been wet that year, and the heaviest machines kept getting bogged down in soft gumbo. But the smaller, lighter tractors, Ford noted, kept moving.

In the course of that year's events in Winnipeg, Ford met Edward Rumely, owner of a company that had specialized in steam engines.

That year, however, Rumely had introduced his first internal-combustion tractor. Nicknamed Kerosene Annie, the new Rumely machine weighed more than twelve tons, moved at a maximum speed of 1.9 miles per hour, and cost a small fortune ($3,400 cash). Ford approved of Rumely's move to internal combustion. Privately, however, he believed that there was no future in tractors that were as large and expensive as Kerosene Annie. He returned to Detroit pondering a machine that would be smaller, lighter, and more affordable—a tractor for the common farmer.

Ford was already well on his way to becoming the flivver king. He had introduced the Model T in 1908, and in 1909 he announced that the Ford Motor Company would make only the Model T to the exclusion of other models. Ford concentrated on making the manufacturing process as efficient as possible so he could cut the cost of the car. His success at this made him both a wealthy man and a populist legend. The Model T cost $900 when it first came out. By 1911 the price had gone down to $780, and the next year it fell to $690. After Ford built the world's first assembly line, the price of the Model T dropped further, to $600, then $550, then $440. Just before the United States entered World War I, the Model T was selling for $360. Ford was underselling every other comparable car by hundreds of dollars. A million cars a year rolled off Ford's assembly line, yet even this wasn't enough to keep up with the demand.

The years from 1910 to 1914 were prosperous ones for America and, for once, farmers fully shared in the prosperity. Many farm families accumulated a little extra cash for the first time. They began to participate, along with the rising urban middle class, in the new American passion for mass-produced goods. Farm families bought Edison phonographs. They bought telephone sets and hooked up to one of the 32,000 rural telephone systems that were operating in 1912. They bought clothing, kitchenware, and a flood of other manufactured articles from the mail-order houses of Montgomery Ward and Sears Roebuck. From local implement dealers they bought new plows, harrows, planters, and binders.

What they did *not* buy, however, was large numbers of tractors. Tractors were still being made the old-fashioned way, one at a time, in plants that looked more like blacksmith shops than modern, Ford-style factories. Fewer than 1 in 500 farmers owned a tractor before World War I. For most farmers, tractors were still too impractical and expensive. They reserved their savings for a different machine: They bought automobiles.

The car was one of the most revolutionary inventions ever to hit rural America. It annihilated distance, relieved isolation, and closed the gap between the disparate cultures of town and country. Automobiles were a particular relief to rural women, whose lives were often more unvaried, tedious, and lonely than their husbands'. Now, for an investment of a few hundred dollars, they could get off the farm, take produce to town, see friends, and do the family shopping. There was some anti-auto sentiment in the country, mostly because of the alarming effect that noisy, reckless cars could have on horses. For the most part, however, rural people embraced the automobile. In 1909 a writer for *Collier's* magazine estimated that the rate of car ownership in New York City was 1 in 190, while among Iowa farm families it was 1 in 34. "Trade your doctor bills for an automobile," went the sales pitch from one car manufacturer, and the prescription worked.

Before 1917 Ford Model Ts probably did more work on farms than tractors did. They pulled plows and planters, mowers and reapers. With the back end jacked up, the rear wheels became belt pulleys capable of powering small threshing machines, sawmills, gristmills, grinders, or corn shellers. Many farmers first acquired a taste and talent for fixing gasoline engines and power machinery not from working on any piece of farm equipment but from tinkering with Tin Lizzie, the family car.

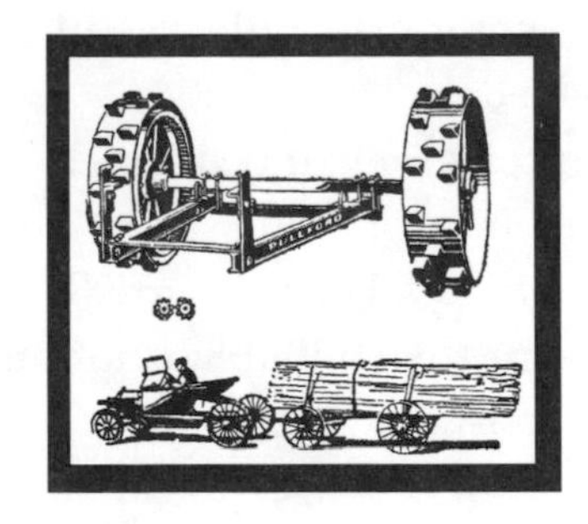

The Model-T did more work on American farms than most real tractors did, until the Fordson came along and changed everything.
(American Society of Agricultural Engineers)

Rumors that Henry Ford was about to enter the tractor business first began circulating during the summer of 1914. His engineers had experimented with a tractor that had a Model T engine and a modified automobile frame. The Model T tractor was nimble, but it lacked the power and traction needed to pull a single fifteen-inch plow: its wheels merely spun in the dirt. That was the hazard of a tractor that was too small and light. Ford's engineers turned their attention to something a little larger but still considerably smaller and more lithe than Kerosene Annie and her cousins.

The demand for tractors in the United States turned upward sharply in 1915 as predictions of a wartime food shortage in Europe came true. Henry Ford, a pacifist, was not above taking advantage of the war to drum up advance interest in his tractor. An early master of the media leak, he let slip bits of news that almost always blossomed into flattering headlines. The Toledo *Blade,* for example, ran an article on May 18, 1915, stating that Ford would soon release a farm tractor that would cost $200 and do the work of six horses, thereby revolutionizing agriculture. The Grand Rapids *Herald* predicted, also in 1915, that Ford's tractor would prove its maker the creative equal of Edison. That autumn, the New York *Call* proclaimed: "FORD'S TRACTOR WILL MAKE WORK OF SOIL TILLER ONE GRAND SWEET SONG." Ford sent a motion picture of one of his tractor prototypes to President Woodrow Wilson, along with a request that federal funds be allocated to deepen the channel of the River Rouge so Ford could, when he was ready, ship boatloads of tractors to war-torn Europe. In 1916 Ford stopped in Kansas City and announced that he would build a car, truck, and tractor, and he would sell all three for $600—"if I don't croak first," he added.

After a long, unexplained delay, the first Ford tractors finally appeared in 1917, in England. "I don't care a damn about your old war in England," Ford reportedly told Baron Northcliffe, Britain's war envoy in the United States, "but I want to help put English agriculture on its feet."

Ford was obliged to call his tractor the Fordson, since a small, opportunistic company in Minnesota had already trademarked the name Ford for its own tractors. The first Fordson for the American market came off a Detroit assembly line on April 23, 1918, less than seven months before the war ended. Henry Ford sent it to his friend Luther Burbank as a gift. When the tractor arrived at Burbank's home in Santa Rosa, California, the famous botanist observed that the machine was "just like Ford, all motor and no frame."

Burbank's quip was true: The Fordson lacked a frame in the usual sense. Instead, the engine itself served as the tractor's backbone. This innovation saved weight (and iron) and simplified the machine's construction. The whole tractor consisted of three segments that were simply bolted together during final assembly. By the end of 1918, Fordson tractors were rolling off the assembly line at the rate of 300 finished units per day, which no other tractor manufacturer could begin to approach.

Fordson tractor, 1917. (U.S. Department of Agriculture)

Tractors before the Fordson displayed a wild variety in form. Most early models were iron behemoths, modeled after their steam predecessors. Their designs were entirely utilitarian, betraying no concern for refinement of appearance. They were essentially heavy iron frames fitted with enormous iron wheels and exposed engines that wheezed, popped, and growled. Some had no seat at all; their operators either stood or climbed around on the frame, making adjustments to the finicky engines. On other tractors, a stamped-out metal dish was stuck on the back for the operator's use, apparently as an afterthought.

Various companies tried to introduce smaller tractor models that would appeal to a wider segment of the farming public, but none of these early attempts turned out to be very practical or popular. Three wheels, four wheels, five wheels; big wheels in the back, big wheels in the front, one big wheel in the center—tractor designers tried every possible configuration, and some that were not so possible. The only thing all of them had in common was that they looked like clunky, graceless relics from the steam age.

The Fordson was different. It was a jaunty little tractor, one that Old McDonald would look right at home on today. Painted gray, it had two big wheels in back, two smaller wheels in front, and a curved cowling above the engine that added a stylish touch and served as the fuel tank. It had a four-cylinder, twenty-horsepower engine that burned kerosene, which was cheaper than gas. Compared with other tractors, it was compact and agile. After 1917, when the Fordson was introduced, the chaos in tractor design began to subside. Tractors began to look increasingly alike, and what they looked like was the Fordson.

Soon it was the best-selling tractor in America. By 1920, after less than two years of production, the Fordson company claimed it had sold 100,000 units—nearly twice the total number of tractors that had existed when Ford's tractor first came out. Ford executives bragged in the early 1920s that they had made more than half the tractors in America, and they were probably right. The initial price was $750, far higher than the $200 that Ford had originally predicted, but still less than any comparable tractor. And, like the Model T, the Fordson became more and more affordable. By 1922 Fordsons were selling for $625, and Henry Ford was making more than half of the new tractors in America. Still, sales declined in the early twenties because of a postwar agricultural depression that drove many farmers and tractor makers out of business. Ford's response to the economic crisis was to cut the price of his tractor by $230, down to $395.

Ford's price cut came as terrible news to the rest of the tractor industry. Alexander Legge, general manager of International Harvester, was out visiting a plant in Springfield, Illinois, when word of Ford's announcement reached company headquarters in Chicago. Informed that he had an urgent call from Chicago, Legge went to the phone, listened for a moment, then erupted, "What? What's that? How much? *Two hundred and thirty dollars?* Well, I'll be. What'll we do about it? *Do?* Why, damn it all—meet him, of course! We're going to stay in the tractor business. Yes, cut two hundred and thirty dollars. . . . And say, listen, make it good! We'll throw in a plow as well!"

International Harvester, Cyrus McCormick's corporate descendant, had been the largest tractor and implement maker in America until Ford came along and knocked the older company into second place. Ford's strategy had been to bring out a small, well-engineered tractor, then leave the product alone while tinkering constantly with the manufacturing process in a relentless drive to lower the unit cost, just as he had done with the Model T. The strategy worked spectacularly well—for a while. Henry Ford awoke the fledgling tractor industry to the need for smaller, more affordable and practical tractors. He also dragged the conservative farm-equipment industry into the age of modern assembly-line production.

International Harvester absorbed these lessons, as it knew it had to in order to survive the economic downturn of the twenties. After matching Ford's price cut in 1922, International lost money on every tractor it sold. Even so, it invested heavily in updating its factories. And it also did something that Ford didn't do—it increased spending on research and development.

Already, in 1918, International had introduced the first truly workable power takeoff (PTO) mechanism. The PTO was a shaft that stuck out of the rear end of a tractor and rotated under power from the engine. This shaft could be used to transmit power directly from the tractor to implements that had previously been ground-driven, such as mowers, reapers, binders, and balers. The PTO was the first feature of the tractor that gave it an indisputable advantage over the horse. (Even the staunchest advocates of horse-farming had to admit that horses didn't have PTO shafts.) By the mid-twenties nearly every tractor maker offered at least one model equipped with a PTO. The one notable exception was Ford.

The PTO improved the versatility and appeal of tractors, but there was still one major farming operation that tractors couldn't handle: the cultivation of row crops. For two or three months after planting, farmers in the cotton and corn belts spent most of their time cultivating. If they didn't, weeds would take over the fields and smother the crops. Corn farmers cultivated their fields four times a season if they could.

Tractors, cumbersome as they were, lacked the finesse needed for cultivation. Even a smaller, relatively lightweight tractor like the Fordson was too broad, squat, and clumsy to do the job. In a field of young corn plants, a Fordson would have the same effect as a bull in a china shop.

Several companies, including International Harvester, came out with stripped-down mini-tractors that they called "motor cultivators." These were spindly, minimalist contraptions—essentially motorized sulkies. The only thing a motor cultivator could do, however, was cultivate. Few farmers could justify owning a machine that cost several hundred dollars and had only one function. Motor cultivators were too specialized. What was needed was a single tractor that could be used for every essential farming operation, from plowing and planting through cultivation and harvest.

In 1921 Alexander Legge gave the go-ahead for an aggressive secret campaign to design just such an all-purpose tractor. It would be tall and maneuverable enough to cultivate row crops, yet it would still have the power to plow, pull PTO-driven implements, and deliver belt power to threshers, grinders, and crop elevators. Engineers in the company's McCormick-Deering division worked on the project for years, building prototypes and filing for patents. They began referring to their new multiuse machine as the "Farmall."

With the introduction of the Farmall in 1924, "power farming" came of age. (American Society of Agricultural Engineers)

Cautiously, without fanfare, International sold its first Farmall to an Iowa corn grower in 1924. Farmalls cost somewhat more than Fordsons, but they could do more. They were the first tractors that could truly put the horse out of work. By 1925 Farmalls were in such demand that International Harvester sold as many as it could make without bothering to advertise.

Now it was the Fordson that couldn't compete. Ford's refusal to improve his product, and not just his production methods, took its inevitable toll. Sales fell off, and Henry Ford was deposed as the tractor king. In 1928 Ford stopped making Fordsons in the United States. By that time, International Harvester was turning out 2,000 Farmalls a month at a refurbished plant in Rock Island. The new Farmall Works was an antiquated farm-machinery factory that Alexander Legge had bought and retooled with a Ford-style assembly line.

THE ALL-PURPOSE TRACTOR and hybrid corn were both introduced in the middle to late 1920s—an inauspicious time to try selling expensive new technologies to American farmers. The loss of foreign markets after World War I (because of newly erected tariff walls in Europe) had dealt American agriculture a grave blow. Farmers were producing more food and fiber than the home market could absorb, so commodity prices took a deep dive. Most industries would respond to this situation by reducing production to push prices back up. Such a coordinated response is possible in industries that have many customers and relatively few producers. In agriculture, however, the situation is nearly reversed: There are many producers (farmers) and relatively few customers (the prime examples being the "big five" international grain merchants and the "big three" meat packers). Individual farmers, when confronted by reduced demand and low prices, have little choice but to maintain or increase their output. This, of course, causes prices to drop even further.

Henry A. Wallace once described this situation, classically referred to as "the farm problem," in this way: "In agriculture," he said, "supply sets the price. In industry, price sets the supply." The result is a structure of consumer prices in which agriculture rather than urban industry tends to absorb the losses when prices fall. "The price structure we have let grow here," as Wallace put it, "is half of steel and half of putty." Farmers were the putty, as they felt the squeeze early and hard during downward phases of the business cycle.

As a result, the 1920s (like the 1980s) roared in the cities, where people could buy an increasing amount of farm products with their gold-based dollars. But the decade fizzled in the country, where the farmer's buying power steadily eroded. Dollars earned from cheap wheat, corn, meat, and milk bought only half as many city products as they had before the war. By 1929, however, even the steel portion of the steel-and-putty price structure could not hold. Pressures that had been squeezing farmers for eight years hit the urban economy—with a resounding crash. The beginning of the Great Depression is usually pegged to the Wall Street debacle of October 1929, but farmers had seen it coming since 1921.

During the late twenties and early thirties, many farmers went to great lengths to buy corn hybrids and tractors in a desperate bid to increase their production. Hybrids gained their first footholds in Illinois and Iowa, where most of the breeders were based and where their products performed especially well. But millions of farmers who would have loved to try the new corn simply couldn't afford it. "When corn sells for ten to fifteen cents a bushel," admitted one Illinois Extension Service worker, "it is rather difficult to convince a farmer that he can afford to take time to become more efficient in his corn production."

It was the same with all-purpose tractors: Most farmers didn't have enough money to buy them in the twenties and thirties. They could look—perhaps longingly—but not touch. The sluggish pace of farm-equipment sales improved briefly in the late twenties, then plunged into an abyss. At Deere & Company, for example, total sales went from a lively $76 million in 1929 to a funereal $8.7 million in 1932. "With everything so quiet it is very hard to get any enthusiasm in work," wrote Theo Brown, the head of Deere's research department, in 1931. "An experimental man needs enthusiasm to do good work. And I can see how this depression affects original thinking."

Among business leaders, the temptation during hard times is to hunker down, cut costs to the bone, and hibernate until it's over. But Brown and other enlightened leaders at John Deere understood how important it was to resist the hibernation impulse and to keep the fires of innovation burning.

During the darkest days of the Great Depression, Brown led a team of experimental engineers in designing two new tractors—the John Deere Model A, introduced in 1934, and the B, introduced a year later. They were both small, economical all-purpose tractors—simple, with good visibility from the driver's seat, a centered PTO shaft, and (an industry first) a hydraulic lift system to raise and lower mounted implements. Farmers fell in love with both models, even if they could only look. By the thousands, they began saving for the day when they could buy one and, as their circumstances improved, they did. The John Deere A and B, produced until 1952, became the most popular tractors in company history, setting sales records that stand to this day.

The Depression years saw improvements in farm machines other than the tractor. Many of the changes were, in fact, made possible by the tractor's newfound versatility. Mechanical corn-pickers, for example, had not been practical before the PTO shaft liberated them from stubborn ground wheels and gear drives. The same was true of combines. They were still ground-driven, large, ungainly, and (for most farmers) utterly unaffordable. Then, in 1930, two Californians went east to try to sell manufacturers a new machine they had developed. The men were Robert Fleming, a farmer-inventor, and his partner Guy Hall, an entrepreneur—the 1930s counterparts of Mark Underwood and Ralph Lagergren. Fleming and Hall were promoting a curious thing: the smallest combine ever built in North America. They called it a "baby combine," with its five-foot cut and a threshing cylinder just as wide. Fleming had designed his baby to be pulled behind a small tractor and powered by the tractor's PTO. The Allis-Chalmers Company paid the two Californians $25,000 for manufacturing rights to the machine. The company made refinements and introduced what it called the All-Crop Harvester in 1934. Allis-Chalmers sold 550 All-Crops in 1935, 8,000 the next year, and 11,500 in 1937—a 2,000 percent increase in three years.

The combine had arrived in the corn belt—this time to stay.

For Elvin F. Canine, my grandfather, 1943 was both one of the best and one of the worst years in his life as a farmer. On the one hand, the small Iowa farm where he and my grandmother, Clara Perkins Canine, lived with their three children was thriving as never before. The 1942 harvest had gone well, and prospects for the current year's crop looked even better. Prices for the milk, meat, grain, and eggs that furnished their income were rising along with the demand for food in Europe, where World War II was turning more and more farm fields into battlefields. Partly because of the war, my grandparents were able to pay off the debt on their farm, which they had bought in 1932 for $125 an acre. The neighborhood consensus was that a farm purchased at that price could never pay for itself, but by 1943 it *had* paid for itself—and in only eleven years. This was cause for celebration.

My grandparents didn't feel like celebrating, though, because 1943 was also a year of sadness and deep change in their lives. My grandfather was a somewhat frail fifty-five-year-old, no longer able to handle

A family photo taken in 1943, just before Wayne (front and center) reported for basic training. Elvin Canine (the author's grandfather, lower right) bought his first tractor a few months later.

the backbreaking demands of farming by himself. He had long relied on the help of his older son, Wayne. By the fall of 1943, however, Wayne was no longer around to help. He was in basic training at Camp Carson, Colorado.

Wayne's departure had been put off for as long as possible, as I learned from a document that I found recently in the attic of the farmhouse where Wayne spent his teenage years and where I, his brother's son, now live. The document is an affidavit, composed with the help of a small-town lawyer.

AFFIDAVIT

I, Wayne Perkins Canine, on oath do depose and say that I am a resident of Warren County, Iowa, and have been classed 1-A in the Selective Service of the United States. That I have rented 40 acres of land adjoining the land belonging to my father, who has 96 acres. That I rented the said land prior to my registration and I am asking that I be deferred for a period of not less than two months and as much longer as the Board would grant me so that I could remain at home with my father and assist with the farming on the 136 acres. That I am at present making my home with my father and mother and a younger brother of 13 years of age.

That my father has no other person who could assist in the operation of the farm, and that we farm all of the land by horses, having no mechanized farming implements.

That I am putting in 24 acres of corn and beans on the 40 acres, leaving 16 acres for pasture. That I am assisting my father on the 96 acres, farming as follows: 30 acres of corn, 15 acres of oats, 8 acres of alfalfa, 15 acres of clover, 13 acres of beans and 15 acres in barn lots and pasture.

That it is difficult to secure anyone to take my place on the farm. That I have had several offers to work on a farm away from home, and in our particular neighborhood there are no available men to work on the farms.

That my father does considerable work on the farm, but is unable to handle the entire farm alone. We raise 60 head of hogs, 8 cattle, 5 horses and 450 chickens.

That I am finding no criticism of my classification and I want to do my full share in the service, but feel that I should be deferred for a sufficient length of time to get in the crops and harvest the small grain and hay.

I make this affidavit in good faith and not for the purpose of avoiding military service.

Subscribed and sworn this 21st of May, 1942.

During the spring and summer of 1943, my grandfather got by as well as he could without his hardworking older son. Plowing, disking, planting, and cultivating with a team were nearly too much for him. He hired younger neighbors to do some jobs that he knew were too strenuous. One such neighbor was Fred Caruthers, a progressive, technically minded farmer. Caruthers not only owned a gas tractor but also an Allis-Chalmers All-Crop combine, one of the first combines ever seen in that part of Iowa. In August of 1943, Caruthers used his All-Crop to combine twelve acres of oats for my grandfather, charging $3.25 an acre—a great bargain and, my grandfather thought, an amazing demonstration of mechanical power.

The time had come, he realized, to harness some of that mechanical power for himself. So on October 22, 1943, he drove to an implement dealer in Des Moines and bought his first tractor—a John Deere Model B. He also bought a tractor plow and cultivator, all for $1,309.12. The new equipment was no substitute for his son, but it would allow an aging farmer to work his place on his own.

In this respect Elvin Canine was exemplary of a national trend. The financial recovery that began with World War II unleashed a flood of demand for farm equipment that had been building for twenty years. The war also removed some six million people from the farm-labor pool (a million who went to war and five million more who left to take jobs in the cities), triggering the most concentrated period of technological adoption in the history of American agriculture. In spite of wartime production limits, the number of tractors on American farms doubled between 1940 and 1945, then nearly doubled again during the following decade—even as the total number of farms began to drop with the rural exodus.

V for Victory: World War II sparked the most concentrated period of farm mechanization in history. (Deere & Co.)

Just as it unleashed a demand for tractors, corn pickers, and combines, the war hastened the advance of hybrid corn. Less than a quarter of the nation's corn acreage was planted with hybrids in 1939; by V-J Day in 1945, that figure exceeded 90 percent.

My grandfather was typical of the hybrid revolution as well. He bought his first bushel bag of hybrid seed corn in 1938 for $6.50. (By comparison, the open-pollinated seed he bought that year cost $2 a bushel.) He experimented with the new corn for a few years, comparing it with the old Leaming strain that he had won prizes with at corn shows since the 1920s. The comparison must have been persuasive, because by 1945 he was planting hybrid corn exclusively—mostly Pioneer Hi-Bred. Leaming was in the same boat as Betty, Doll, Rex, Daisy, Edna, and a dozen other workhorses that had served my grandfather well and faithfully over the years. "Power farming" and hybrids had suddenly turned them all into relics.

My grandfather never felt completely at home in the postwar world of hybrid corn and power farming on a large scale. Born in 1888, the same year as Henry Wallace, he had come of age in the era of corn shows, hand-husking contests, and the small diversified family farm. Now he listened as Ezra Taft Benson, President Eisenhower's secretary of agriculture, told farmers they should "get big or get out." So in the late 1950s he got out. He, too, had become a relic. Technology had made my grandfather obsolete.

THE ENTREPRENEUR

Two or three months of hard hustling among capitalists had shown me a little of the disparity between the inventor's and the investor's points of view.

—Lee de Forest (1950)

Well, it's all gone, but we had a hell of a good time spending it.

—Thomas Edison, after losing $4 million on a new magnetic iron ore–separation plant (c. 1900)

On a list of the nation's financially and socially elite ZIP codes (Grosse Pointe 48236, Shaker Heights 44120, Beverly Hills 90210), Lincoln 67455 does not appear. Lincoln, Kansas, is the town where Ralph Lagergren grew up. It is the seat of Lincoln County, a place known not for its millionaires, of which there are none, but for its limestone fence posts, of which there are several thousand. Three-foot-tall limestone obelisks outline the farmsteads and fields of Lincoln County like rows of miniature Washington monuments. Ralph's father worked in the county office of the Soil Conservation Service; his mother taught school. His background did not endow him with ties, familial or social, with the sorts of people who attend hunt balls, charity auctions, and $1,000-a-plate political fund-raisers. So when he started trying to raise money for the combine venture, he found himself in unfamiliar territory. To seek advice on raising capital, he went to the Small Business Administration and a few similar agencies. He was given some brochures, pamphlets, and books, all of which he found utterly unhelpful. What the brochures and booklets all said, essentially, was that he should approach either wealthy friends and family members ("I have one aunt with a little money, but she said no") or venture capitalists.

Venture capitalists, Ralph soon learned, had no interest in the cousins and their combine. In general, venture capitalists demand large ownership stakes in the development of sexy, high-tech products. Agriculture, in their view, was not sexy. What venture capitalist dreams of sitting in a Ferrari and chatting on the cellular phone about a combine? Besides, venture capitalists, Ralph learned, did not venture their capital on untried inventions; that would be too risky. They looked for promising new products that had already been through the gauntlet of conceptual development and were ready to hit the marketplace. The Bi-Rotor combine was far from that point. The fully realized design existed, as yet, only in Mark Underwood's head. When Ralph first went looking for funds, all he had to show prospective investors was a couple of patents and Mark's drill-operated model.

Since Ralph had no wealthy friends or social connections, he decided to make some. The Dallas–Fort Worth area, where he lived, was a good place to do this. It had millionaires by the mile. They were lined up like limestone fence posts in neighborhoods such as Highland Park, Stevens Park, and Kessler Park in Dallas, or Rivercrest, Westover Hills, and Sundance Square in Fort Worth. They had well-fenced ranch spreads north of town, where oil and cow money parked its longhorn Lincoln Continentals and pastured its pedigreed horses. Oil fortunes weren't what they used to be, but the money still flowed like wildcat gushers in Dallas and Fort Worth society.

Ralph felt most comfortable in the downstairs milieu of the help—the social register's personal secretaries, chauffeurs, and helicopter pilots. Just as he had once courted the favor of nurses in order to influence the doctors they worked for, he now flirted with secretaries to gain privileged access to their bosses. He developed a kind of high/low strategy, working the upstairs and downstairs crowds simultaneously. He wrote frequent, ingratiating letters, for example, directly to Ross Perot and Sam Walton (two early prospects), while also befriending a few key people on Perot's and Walton's personal staffs. He avoided anyone in the middle—vice presidents, middle managers, and other lieutenants, who would neither help him gain access to the boss nor make risky decisions on their own.

He angled for the big fish first. Besides Perot and Walton, he went after names like Bass, Getty, Pickens, and Hammer. "I shamelessly put myself forward," Ralph told me. "I went around to places where I might meet those kinds of people. I talked to people, who gave me the names of other people, and I followed up on every lead."

Here is a typical episode from this phase of Ralph's fund-raising efforts. "I got to know one of Ed Bass's people a little bit," Ralph told me. "I persuaded her to get me an invitation to a party Ed was throwing. It was held in a fancy party room at the Worthington Hotel in Dallas. I was talking to people and having some good conversations, working my way around the room. Finally somebody said, 'Who *are* you?' But they didn't kick me out, so I guess I won them over. By the end of the evening, people were shouting across the room, 'Good night, Ralph—and good luck with that combine!' "

He met many interesting people and sampled some good hors d'oeuvres this way, but he failed to raise any capital. "I wasted too much time trying to track down the big-, big-money people," he said. "They get hit up a hundred times a day. They delegate most of their investment decisions to professional money managers"—lieutenants, in other words, who were paid to be cautious. The minute he was referred to one of them, he knew it was all over. So he decided to aim a few rungs down on the social register, targeting prospects who had plenty of money but were not famous for it. He tried to identify people who had enough money to be able to take a $25,000 flier on a high-risk venture with no sweat but weren't so wealthy or well known that their every movement showed up on the radar screens of yacht salesmen and gossip columnists all over America.

He also looked for people who had some interest in agriculture—who at least knew what a combine was and how it worked without the need for lengthy explanations. He got his first break with Roger Carpenter, a successful entrepreneur and investor who lives in Coffeyville, Kansas. Carpenter grew up on a cattle farm in Oklahoma. By a roundabout route that landed him in various classrooms as a teacher for a while, he eventually wound up "in the hamburger business," as he puts it. He became a fast-food mogul, counting among his properties a number of Sonic drive-in restaurants in the southern Plains region. He met Ralph through Scott Larson, a lawyer in Dallas whom Ralph had persuaded to provide free legal counsel in exchange for a stake in the combine's future earnings. With the same guileless sales pitch that had convinced Larson of the Bi-Rotor's virtues, Ralph put the hooks into Roger Carpenter.

"He asked me if I knew anything about combines," Carpenter told me one day in a deep, relaxed voice with a slight Oklahoma drawl. "I said, 'Enough to be dangerous.' We just hit it off. I listened to him describe Mark's ideas, and I decided they knew what they were talking about. But I knew they had a hard row to hoe. When you have a good idea but no capital to produce it, it's like trying to fight a war with no generals. I became their original investor, I guess. But more important, I got Ralph an audience with various other people who could help."

It was Carpenter who first introduced Ralph to the people at KTEC, for example—a connection without which Ralph doubts he and Mark would have gotten the combine off the ground. Carpenter also advised the cousins to form a limited partnership, which they called Agri-Technology L.P., and introduced Ralph to several of the people who eventually put up $25,000 each to buy into the partnership. For that amount, investors would receive a 2.5 percent share of any royalties the Bi-Rotor combine earned, if it ever earned a penny.

One day I listened while Ralph talked to a prospective investor on the phone. "The last major combine innovation came out in the late seventies, when International introduced the Axial-Flow models," Ralph said into the phone. "It took the company one hundred and twenty-five million dollars to bring it to market. Ours is the first major innovation to come along in fifteen years. We're trying to bring it along far enough so one of the big companies can pick it up for less R&D investment than it would take them to develop a major innovation by themselves. We're well on our way to that goal. At this stage the shares are going for twenty-five thousand for two and a half percent of the deal. Can I stop by next week sometime?"

Ralph made dozens of calls like this every day. He crisscrossed the Plains in his red GMC pickup, calling on leaners who looked as if they might fall his way. One by one they fell, until he had a dozen. They included some well-to-do farmers in Texas and Kansas, a South Dakota banker, a Kansas physician, a Texas lawyer, the owners of a seed company, a machinist turned entrepreneur, and a second fast-food millionaire. "I got the best group of investors anybody could ever hope for," Ralph told me with Texas gusto.

Thus funded, the cousins built Whitey and ran her for a summer.

The machine performed better than they had hoped it would. Still, no major manufacturer stepped forward with an offer to license the technology. That's when Ralph and Mark realized that they would have to take the entire project a major step further. They would have to build another machine, one that embodied Mark's ideas from the ground up. They resolved, if necessary, to take the machine all the way to the manufacturing stage by themselves.

By the beginning of 1992, when they started work in Haven on what they called the "ground-up" prototype ("It sounds like hamburger, doesn't it?" Ralph laughed), they had exhausted most of their funding from the limited partnership. They would be running on financial fumes by summer. To postpone the day of reckoning, Ralph rounded up a few more investors, bringing the number of limited partners to fifteen. He had also committed shares of the Bi-Rotor's future earnings to various people who had helped out along the way, such as the owner of the Texas machine shop where Whitey was built. That left Mark and Ralph with just under 50 percent of the deal for themselves. Ralph couldn't just keep selling shares to more limited partners or he and Mark would eventually be left with nothing. Yet if they didn't tie into some more money soon, the whole dream would come crashing down.

Ralph had not expected to approach major manufacturers with formal proposals until after the ground-up prototype was finished. Then, assuming the machine proved everything it was supposed to prove, he would have something truly impressive and valuable to sell. At that point he could negotiate from a position of strength. But now, with the wolf at the door, he changed his strategy. He came up with his two-million-dollar plan. For that sum a company could buy the first option to market Bi-Rotor combines. The money would pay for three to five years of research and development, which would include the construction of a half-dozen prototypes, culminating in Ralph's ultimate marketing vision—the "Bi-Rotor Brigade," a triumphant sweep of Custom-Cutters' Alley, from Texas to Canada, with a fleet of revolutionary new combines.

"I know two million dollars sounds like a lot of money," Ralph told me. "But it's not extravagant considering all we'd give for it. Hell, John Deere spends a couple million dollars on R&D in one week. Whoever does this deal with us will be getting a market-ready machine for a fraction of what it would cost them to develop one in-house.

"I have an inner, instinctive time clock, kind of like an animal that knows when it's time to mate," he added. "It's time for us to mate with a big company that can see us through this thing."

The list of potential mates was rather short. The obvious ones could be counted on a farmer's hand, which, like a carpenter's, often lacks a digit or two. There were really only two: John Deere and Case-International Harvester—Big Green and Big Red, who together accounted for something like two-thirds of all American combine sales. Ralph sent copies of his two-million-dollar proposal to both companies.

He had his hopes pinned on Case-IH. The people in Racine, Wisconsin, where the company is headquartered, seemed curious and receptive, whereas John Deere had so far remained indifferent. Ralph had done everything he could to fan the flames of desire at Case-IH. He had cultivated an inside source, a manager in the harvester division in Racine. The man, whom I will call Chuck Haldeman, was a member of the Case-IH delegation that had come down to inspect Whitey the previous fall. Ever since, Haldeman had been enthusiastic about the Bi-Rotor and its prospects. "If he ran the company," Ralph said, "we'd have a deal."

Haldeman, however, did not run the company, and no deal had materialized. "We aren't taking you guys lightly," Haldeman told Ralph, "but we're struggling with budgets and a tough market right now. We're very interested."

This sort of thing had gone on for more than a year, but the courtship failed to progress beyond flirtatious phone calls. Ralph hoped that his $2 million deal—the first formal proposal that had passed between them—would break the impasse.

Weeks went by with no answer from Case-IH. There was no "yes," no "no," no counterproposal. Ralph decided it was time to go to the top. He wrote a letter to Bob Carlson, then CEO at Case-International, requesting a personal meeting. Carlson didn't respond, so Ralph wrote to Carlson's boss at Tenneco, the parent company. Once again, there was no reply.

In the spring of 1992, Ralph concluded that his pursuit of Case-IH wasn't going anywhere. He decided that he'd better start playing the other side of the street. This time he sent letters to each person on the board of directors of Deere & Company. In the letters he explained that the Bi-Rotor combine represented unique advances in the technology of harvesting, advances that Deere & Company should at least consider adopting but which it had so far ignored. Was it really in the best interests of the company to remain uninformed about an innovation that could one day challenge its market share in combines?

Not long after that, Ralph got a few calls and letters from John Deere people—"just little sniffs of interest," he said. The stirrings didn't go much beyond a tepid acknowledgment of the Bi-Rotor's existence. So Ralph ratcheted up the pressure in the only way he knew how. He sent out yet another batch of letters, this time to a dozen or so of John Deere's major institutional stockholders. "I just want a few big stockholders to get on the phone to somebody at the company and say, 'Why aren't we checking out these guys?' " he explained.

While Ralph was courting the Big Two, progress on the prototype was painfully slow. Mark and Ralph had seriously underestimated how long it would take to design and build a completely new combine. They had built the Kansas State prototype in forty days and Whitey in four months, but the new prototype was a job of another order of magnitude. Mark was designing and building an entire combine virtually by himself, piece by piece. Building a three-bedroom house single-handedly would have been far easier. Mark made detailed drawings of each part on a drafting board in the Winnebago, then handed the drawings to machinists in the Kincaid shop who fabricated the parts out of metal stock. The cousins paid for fabrication labor by the hour, with the result that the stack of labor invoices grew faster than the combine did.

Ralph had told his investors that he hoped the new machine would be ready in time for the 1992 wheat harvest. As June approached, however, the wheat matured and the prototype did not. There was no way it would be ready to cut wheat. It was no more than a bare steel skeleton on wheels, with no engine, no cab, no guts of any kind.

Making the best of the situation, the cousins suspended their work at Kincaid, loaded Whitey on the combine trailer and trucked it up to the Underwood farm near Burr Oak. Mark would cut wheat, testing the self-leveling sieves and other systems, while Ralph addressed the money problem.

"This is when most people fail, is right now, when money starts to run out and your time and energy level start to fade," Ralph told me one day that summer. We were sitting in his pickup on a gravel road, watching while Mark cut wheat. "It's going to get real tough. One thing I've got to do is keep the Bi-Rotor visible in the media. I'm trying to get articles about the sieves into newspapers and magazines, to keep interest in what we're doing alive at a grassroots level."

Mark came by with Whitey and signaled that he was getting low on fuel. Ralph checked the diesel-fuel tank on his cousin's pickup and found it almost empty, so he and I drove to the nearest town to fill up.

"I'm still pushing to get the prototype out in time for the fall crop," he said as we bounced along a narrow dirt road. Wheat fields stretched away to the horizon in every direction. Four John Deere combines, operated by a custom-cutting crew, moved across one field in perfect formation, like National Guard jets on Independence Day. "I know it's a long shot to have this thing in the field by October, but I've got to have goals. Mark is scared that if we rush it and roll out too soon, we'll have problems. There's some merit to that. But we can't just stay in the shop and perfect it forever. I need to keep pushing. That's part of my job."

He told me that the people at KTEC had recently sent him a copy of an article from *Entrepreneur* magazine about the differences between entrepreneurs and inventors. "The article has a picture on the first page of two guys playing tug-o'-war," Ralph said. "One guy is wearing a suit, the other guy is in jeans. As a joke, somebody at KTEC labeled the two guys 'Ralph' and 'Mark.' We thought it was pretty funny. But then I started reading the article, and I was amazed. It really could have been written about me and Mark. We're great partners, but there have been times when I just couldn't understand him. I mean, he doesn't even wear a watch! I've tried to tell him how he could get better organized and he just ignores me. But this article really helped me understand our differences."

Ralph later gave me a copy of the article, written by K. M. Schmidt and Michelle R. Schmidt. "It is one thing to invent something and quite another to commercialize it," the piece begins. "Few entrepreneurs invent anything, and few inventors become successful entrepreneurs. Only a team can do it. Understanding is therefore crucial."

The article includes a chart listing the ways in which inventors and entrepreneurs differ. When Ralph read down the list of traits, he felt as if he were reading horoscope descriptions of himself and his cousin. The descriptions, he thought, seemed uncannily accurate.

INVENTOR	ENTREPRENEUR
• passive	• aggressive
• introverted	• extroverted
• mechanical	• verbal
• highly emotional	• keeps emotions hidden
• competes with himself	• strong desire to compete with others
• nonsocial; feels uncomfortable in groups	• active socially; loves people
• dislikes structure, time limits, and definite plans	• likes structure; highly dependent on schedules and plans
• compulsive perfectionist	• not concerned with doing everything correctly

"Mark and I, I think, are a perfect team," Ralph said as we filled the fuel tank. "Without him I'm nothing, and without me he's nothing. We need each other. But you know something? We can't continue to put in long hours for not much reward forever. It's taking a toll on both of us. If we can hook up with a company, then I can see finishing this up the way we should, and not just working like dogs. I'm hoping to hear something from John Deere any day now."

I was sitting in the kitchen at Mark's house one morning during the last week of June, sipping coffee, when the phone rang. Deb answered. After a moment she put a hand over the receiver and said to Mark, "It's a guy from John Deere." As she handed the phone to him, she gave him a stern look and whispered, "Ralph said if somebody called not to tell them anything." Ralph himself had run to town to get some spare parts for Whitey. He would be sorry he missed this call.

Mark and the Deere man chatted. "We've been cutting wheat down south of here with Whitey," Mark told him. "Conditions have been terrible. I've never seen wheat so bad. They've had hailstorms down there about every Friday night for the last three weeks. It's been a lousy year for farming. But the other way to look at it is that it's been a good year to test Whitey in some tough conditions. She's been working real well. . . . Yup. Okay. Bye."

Mark hung up. "It was one of their combine engineers," he announced. "He wants to come down next Tuesday. Sounds like a pretty decent feller. Not one of them nose-in-the-air types."

The engineer from John Deere arrived on the seventh of July, as expected, and checked into the Dreamliner Motel. Ralph met him in his room to do a little preselling.

"The first thing he said when I met him was, 'We're looking for a smaller machine with high capacity,' " Ralph reported later. "That sounded right up our alley. While we were still in the motel, I showed him a picture of the sieves. He looked at it and said, 'You mean each panel pivots?' I said, 'Yup. But we'll see these later,' and I put the photo back in my briefcase. He just about jumped in after it."

Then Ralph took his visitor to Mark's place, where Whitey was waiting, newly washed. The engineer had no sooner said hello to Mark than he made a beeline for the back of the combine, ducking under the cowling to see the sieves. Mark blew on the sensor to activate the leveling action. As the sieve panels rotated, the visitor said, "Well I'll be darned."

Ralph and Mark looked at each other. They were on the verge of losing their composure and bursting out in laughter. "We've heard that so many times from people we've talked to over the past couple of years," Ralph told me later. "They all say, 'I'll be darned.' It's a sure sign of interest."

Mark opened up the side panel to reveal a portion of the rotor and cage. "You can't run that same hole size on the cage with all crops, can you?" the engineer asked.

"That's what we've been doing, and it works just fine," Mark said.

"Well, I'll be darned."

Later that afternoon, when Ralph was about to take the Deere man back to the Dreamliner, Ralph saw him looking at Whitey and shaking his head slowly.

"What's the matter?" Ralph asked.

"You know," the man said, "it would have taken us millions of dollars to do this." He told Ralph he wanted to figure out a way they could all sign something—a nondisclosure agreement—"so we won't be afraid to talk to each other."

The meeting had accomplished everything Ralph hoped it would. He felt optimistic that an agreement with Deere would follow soon.

The next day, as luck would have it, Ralph got a call from Chuck

Haldeman at Case-IH. Haldeman said, "We're still very interested, and we're trying to put something together for you."

"Well, I have to tell you," Ralph replied, "John Deere was here yesterday."

Haldeman wasn't too happy to hear that. He and Ralph agreed that if Case-IH lost its competitive position in combines and John Deere got a bigger market share than it already had, it would have an adverse effect on farmers. The price of combines would go up even higher, and the pace of innovation would slow down.

"I'd rather go with the underdog—Case-IH—and help keep the farm-machinery industry competitive," Ralph told me. "But I wouldn't turn down John Deere, if that's the way it goes. Hell, we'll do a joint venture with Russia if we have to." (In fact, that summer they showed Whitey briefly to Boris Yeltsin when he visited a Kansas farm, and Yeltsin seemed to understand how the Bi-Rotor worked. He made his hands circle around each other and asked his interpreter, "You mean they both move?"—meaning both the rotor and the grate. Ralph and Mark were impressed.)

Ralph expected to hear from Deere any day. He went so far as to ask Scott Larson to draft a nondisclosure agreement for Deere to sign. The agreement, a fairly routine device in patent negotiations, would have bound both parties to maintain the secrecy of any proprietary information disclosed during the course of exploratory discussions. It would also have obliged both parties to refrain from copying any ideas disclosed by the other side during the course of such discussions.

Scott Larson advised Ralph to make notes on all conversations and meetings with Deere people, detailing everything related to Mark's patents that was said or shown to them. Ralph was to record the date on his notes, then send them to Larson for safekeeping. The notes would serve as evidence in case they ever had to sue Deere for patent infringement.

Ralph got the nondisclosure agreement ready, then waited to hear from his man at Deere & Company. A week went by. Two weeks. A month. No call. Summer turned to fall. In Kansas the corn, milo, and soybeans lost their chlorophyll and turned a russet brown. Soon the grain would be ready to pick. Farmers greased their combines for the fall crop. In early September a call finally came, but its substance was not what Ralph hoped it would be. The caller wondered if Deere could send some people down to watch Whitey during the corn harvest.

Ralph asked if they were prepared to sign a nondisclosure agreement. No, they were not. They just wanted to have another look, to see how Whitey and the sieves handled corn.

Ralph felt that he and Mark were being trifled with. "That's the kind of game they play," he said. "They want to see more, but they're not willing to put out anything in return." He told the caller he thought Deere had seen enough to make a decision of some kind—a decision that implied at least a small degree of commitment. Ralph's response in effect, was, "Thanks for your interest, but we'll pass." The conversation ended there.

Ralph's discussions with Case-IH had also reached a dead end. "They just flat don't have the money," Ralph lamented. Case's parent company, Tenneco, was leaking cash like a grounded oil tanker. Tenneco had reported a loss of $732 million in 1991, mainly because of a $618 million restructuring of its Case division. By the fall of 1992, Case was well on its way to racking up a loss of $228 million for the year. Though this was an improvement over the previous year, things still looked grim. Projections for tractor and combine sales for the coming year were pessimistic, and a debacle at Case looked more and more likely.

Case-IH wasn't the only farm-machinery company that was suffering in the early 1990s. The industry was, at best, in a state of febrile paralysis. When the recession of the nineties hit, agriculture still hadn't recovered completely from the farm crisis of the eighties. It seemed a particularly inopportune time to try to sell a radical new idea to the farm-equipment industry. Ralph didn't see it that way, though. "To me," he said, "this is the *best* time we could be doing what we're doing. There's a huge pent-up demand out there for new combines. Farmers are waiting for new technology that they see as a good value. What they *don't* want to do is fork over a fortune for technology that's been around for years and years. What's out there now is warmed-over leftovers. We've got a new dish to serve up, and farmers are hungry for it."

They had the dish, but no restaurant. In fact, money was getting so tight that they were hard-pressed to feed themselves. They were living mainly on funds borrowed from one of their more sympathetic investors. They had gone back to work at Kincaid for a few months following the wheat harvest, but in September they suspended work on the prototype once again while Ralph searched for white knights who could rescue them. Mark went home again, this time to harvest corn and milo. Then he parked Whitey and spent several weeks doing dozer work just to earn grocery money for his family. When he wasn't combining or dozing, he was in his shop working out bugs in the self-leveling sieves. Also, he finished drawing up patent applications for two more combine innovations: new grain-elevation and unloading systems, simpler than conventional augers and gentler on grain. He would incorporate these ideas into the new prototype—if it ever got built.

Ralph hit the road again, looking for a deus ex machina. He visited Chance, his eight-year-old son, whenever he could dip down into Texas. His base of operations remained at Kincaid, even though the prototype's bare skeleton had been pushed far back into a dim corner of the shop, near the scrap-iron bin. After one of his trips, Ralph returned to Haven to find that the Winnebago—the closest thing he had to a home—was gone. It had been removed from the shop and parked outside. Delmer Kincaid was apologetic, but with work on the prototype at a standstill, he said he needed the space for other projects. Outdoors, the trailer lacked both plumbing and heat; its propane heater was malfunctioning. Ralph spent some chilly nights as close to despair as he had been since the dark time, two years before, when his mother died and he and his wife were divorced. "If I quit and go back to work someplace now, it will die," he said.

By "it," he meant the dream. The dream was the most important thing in his life, after his son—but even Chance (who earned his name by surviving some harrowing complications of birth) was taking a backseat to the combine project. For its sake, Ralph had quit a good job, lost his wife, sold his house, and either sold or stored most of his possessions. He divided his days and nights between his pickup truck, a borrowed office at Kincaid's, and the unheated camping trailer. If the dream died, what would he do?

He never allowed himself to dwell on that prospect. Though not religious in the conventional sense, Ralph possessed a deep and unquenchable faith, a very American faith, in the power of positive thinking. The nugget of his personal gospel was expressed in a quotation from Thoreau's essay, "Civil Disobedience," that Ralph once showed me. "If you advance confidently in the direction of your dreams and endeavor to live the life you have imagined," Thoreau wrote, "you will meet with a success unexpected in common hours."

"If you're prone to depression and don't know how to get yourself out of it, then you should never do something like what we're trying to do," Ralph said. "Sure, I feel low sometimes. What do I do to get out of it? I look at the newspaper. I see stories about wars, people dying, children with fatal diseases. Then I start to feel pretty lucky. I think, 'What have I got to complain about? I'm living my dream.' It takes me less than fifteen minutes, and I'm back up and ready to go."

Gloomy thoughts tended to close in on him whenever he spent evenings sitting around in the camper by himself, so he avoided this whenever possible. Every night he sat in the Kincaid office typing letters or planning his next day's phone calls until 11:30 or so, then he went to bed.

On weekends he went out dancing. He would put on a pair of pressed Wrangler jeans, a colorful western-style shirt, a pair of tan lizard-skin cowboy boots, and no hat ("I've never favored a hat. It just doesn't feel right"). Then he would head for a country-western nightclub like McGraw's in Wichita, or B.G.'s Cadillac 'n Cowboys in Hutchinson. His strongest romantic interest in years—a quick-witted, fun-loving food broker named Kay—had recently been transferred from Kansas City to Louisville. Ralph was saddened but undaunted by this setback in the small sliver of his life that was not taken up by the combine. He had no interest in one-night stands or in playing the field; he was a serial monogamist between ports. He looked for a girlfriend in the same disarmingly straightforward way that he sought mates in the combine venture. He would walk up to a woman and say, "Will you dance with me before a line starts forming here at your table?" Or he would lean over and confide, "My mom is sitting back there in the corner, and it would make her so happy to see her son dance with a pretty girl." Sometimes he paraphrased Al Pacino's character in *The Scent of a Woman:* "Would you mind if I sat here? It would help keep some of these womanizers away."

During the days he pursued about five lines of offense at once. His main goal was still to convince a single major corporation to go for the two-million-dollar plan. He fired off letters to corporate presidents and CEOs and responded to inquiries from farmers. He continued to write and call editors at newspapers and magazines to suggest articles on the combine or on Mark's more recently unveiled self-leveling sieves. A professional publicist could hardly have succeeded better in reaching a target audience. During the 1992 growing season alone, major articles on the Bi-Rotor combine and self-leveling sieves appeared in *Farm Journal, Successful Farming, Farm Industry News, Corn Farmer,*

Kansas Living (published by the Kansas Farm Bureau), *Working Tires* (sent to dealers and commercial customers of the Goodyear company), the *Kansas City Star*, and the Wichita *Eagle*.

The story in the *Eagle* happened to catch the eye of a Wichita resident named Bernard Taylor. Taylor was a manufacturing representative, a matchmaker between people who have things they want to build and other people who have factories where the things could be built. He read the article about Mark and Ralph with intense interest. Here were two guys with a big machine to build, and Taylor knew just where it could be built. He called Directory Information and got Mark's number in Burr Oak. Mark happened to be at home for the corn harvest and picked up the phone when Taylor called.

"If you want somebody to build your combine," Taylor said, "we've got a plant where you could do it."

"That's the kind of thing my cousin handles," Mark replied. He jotted Taylor's number on a scrap of paper, then went out to cut corn with Whitey. He forgot about the phone call until about a week later, when the wadded-up scrap with Taylor's phone number resurfaced. "Oh, yeah," Mark remembered, and passed the number along to Ralph. "The guy sounded serious," Mark reported.

"Thanks a lot," said Ralph. He wished once again that Mark was better organized and made a mental note to buy his cousin a Filofax or something. Then he dialed the number. Like most of the hundreds of leads he had followed up on during the previous year, the call to Bernard Taylor was a shot in the dark.

But this shot hit something solid.

Bernard Taylor was eighty-two years old when he and Ralph first talked. He was a large man with a full head of thin auburn hair and liquid blue eyes that peered alertly through brown horn-rimmed glasses. He always dressed sharply in a tie and jacket. His mind was as keen as it had always been, though his body had begun to fail him in his ninth decade of life. He walked slowly and painfully with a black cane. Every day he went to work in an office suite that he shared with his son, James, who was his business partner.

A native of Kansas City, Bernard Taylor was lured to Wichita in 1928 to play semipro baseball for thirty-five cents an hour. He also worked part-time as a night watchman at the Steerman Aircraft plant, which was soon absorbed into the aeronautics empire of William E. Boeing. Taylor loved baseball, but Boeing offered more stable employment, and the young athlete was discovering new skills as a manufacturing man. After the Depression ended he went to work full-time at Boeing. By the end of World War II, he had risen to production manager in a factory complex that made big planes—fifteen-man gliders and B-29 Superfortresses. But Bernard chafed under the government controls that had gone into effect during the war and persisted, to an extent, after it was over. So he left Boeing in 1949 and started a series of small manufacturing plants of his own. All of his plants made parts and ground-support equipment for the aircraft industry.

"I knew a lot of people in the airplane business," Taylor told me in his office one day. "People at Boeing, Cessna, Beech, and later at Lear-jet. So it was easy to get work, but we worked like hell." He eventually sold his plants at a handsome profit but stayed on in the manufacturing world as a matchmaker. Now he pursued deals with the vigor of a man half his age. You could tell he got a kick out of it, savoring the challenges and the unexpected turns as if he were still a kid playing baseball.

He and Ralph liked each other at once. They were cut from similar cloth, a bolt labeled "Plains entrepreneur." They were both as open and honest as the Kansas prairie, yet also canny and shrewd, appearing to reveal much, but revealing less than they took in. They met in Wichita, and both of them liked what the other had to say. Taylor knew a bit about the farm-machinery industry and also something about combines. Ralph's argument convinced him that the combine business was due for some shaking up. He was also convinced that Mark's reinvented combine—not just the Bi-Rotor threshing system but the total package of inventions that the new prototype would embody—was the machine that could do the shaking. He knew of some manufacturing companies in the Wichita area with defense and aviation contracts that had dried up. The companies needed to diversify their product bases. One of them, he thought, might even be interested in Ralph's two-million-dollar deal.

First they'd have to put together a proper proposal, complete with cost and revenue projections. "How much would it cost to make a Bi-Rotor combine?" Taylor asked. Ralph had no idea. He could sell blame near anything, but manufacturing was beyond his ken.

Taylor wasn't too surprised. He'd seen it all before. But he posed an uncomfortable question: How was Ralph going to convince somebody to risk a couple million bucks if he couldn't point out—in dollars, not just blue-sky salesman's patter—exactly what was in it for them? It was time to work up some numbers.

Bernard Taylor and Ralph spent a long Saturday in the older man's office plugging numbers into a spreadsheet. They filled in the blanks on projected costs, sales, profits, and losses over a five-year period. They estimated the number of man-hours it would take to produce each machine. They guessed what hourly labor rates would be in the years 1995 through 2000. When they came to the blanks for parts-and-materials costs, Taylor called a friend at Learjet who once worked for a farm-equipment company. The friend helped them come up with some credible figures. They sat in front of the computer all day and into the evening.

"A lot of these numbers we pulled out of the air," Taylor said. "You're always afraid when you estimate man-hours and things that you'll be way off, but it's surprising how close you come. I kept all the estimates high, padding our costs so they were on the pessimistic side. We ran our projections out for several years and it was *beautiful*. What amazed me was that we never ran out of money. In the third year of production we were making ten to twelve million dollars, and it kept going on from there. I mean, these were numbers from the dream world! It can't help but be a profitable thing. This has got more potential than anything we've ever been involved in."

The factory where Taylor thought the combines could be built was in southeastern Kansas, on the edge of the Ozark Mountains. Though small by Fortune 500 standards, the manufacturing facility was well equipped with precision metalworking equipment, which it had used until recently to fulfill lucrative defense and aviation contracts. Ten years ago the company made items ranging from small clips for M16 rifles to huge weldments used for hanging nuclear cruise missiles under the wings of B-52 bombers. The company made the cruise-missile hangers for $35,000 each and sold them to Boeing, which made the bombers, for $75,000. Bernard Taylor had procured these contracts. In fact, he had owned the plant, and his son, James, had run it. They sold the plant in the mid-eighties to a Chicago-based holding company known for investing only in sure moneymakers.

"The plant was making beautiful money when we sold it," Taylor told me. "But then the contracts ran out and weren't renewed. Those defense contracts were this plant's bread and butter. The irony is that ever since we sold it, we've been looking for a new product line to bring into the place. This combine would be an ideal thing for it. It could just *make* that plant—turn it from a two-hundred-person operation to a two-thousand-person operation."

The plant's new owner had plenty of money to invest. The Chicago holding company operated sixteen industrial properties with combined sales, in 1991, of $335 million. The holding company was owned, in turn, by a young and successful leveraged-buyout artist in New York. The buyout boss had more than enough capital resources to swing Ralph's two-million-dollar deal. But first Ralph had to sell one of the boss's lieutenants—the president of the holding company—on the idea. Ralph was wary of lieutenants, but he had no choice other than dealing with him. Ralph had neither the time nor the money to go to New York and schmooze with the big boss's secretary and, perhaps, get invited to the buyout artist's birthday party.

Ralph prepared the most elaborate sales presentation he had ever given in his life. He put together a thick, spiral-bound prospectus on the combine, prominently featuring the attractive profit projections that he and Taylor had come up with. He outlined a speech, memorized it, and arranged an accompanying sequence of video clips, slides, charts, graphs, and drawings. "This is the turning point of the project," he told me a few days before the crucial meeting. "This will be the toughest sale I've ever made, but I'm confident it will happen." What he didn't say was that if the deal failed to happen, he might be out on the street looking for a job soon.

On a Tuesday in late September a man whom I will call Rod McGuinn, president of the holding company that bought the Kansas plant from Taylor, arrived in Taylor's office suite. He had flown down to Wichita in a Cessna Citation, a small private jet. Surrounded by Taylor's collection of western art—reproductions of Remington sculptures and Russell paintings of cowboys shooting at one another—Ralph gave his presentation. McGuinn appeared receptive and, at times, impressed. After the meeting broke up and McGuinn flew off in his jet, Taylor congratulated Ralph on a fine job. Ralph was exuberant. He felt certain that he had hooked his fish. "I can finally see this thing working out the way it needs to," he said. "We're getting *so* close."

McGuinn called a few days later. He was cordial and frank. "What we do very well is buy established companies with growth potential," he told Ralph. "We keep existing management in place and help them with cash and strategic planning. What you guys are trying to do doesn't fit in with that. Sorry. I'm sending you a letter, but I wanted to call first to let you know."

Ralph couldn't believe it. He had done his best and lost the sale. Now what? Taylor knew of several other plants in Kansas that could help build the prototype. But none of the others had access to the kind of capital that McGuinn did. The two-million-dollar plan, which had seemed so reasonable, perhaps even inevitable, to Ralph when he had formulated it, now began to look like a remote fantasy.

Ralph had not often felt panic before in his life. His unquenchable can-do attitude had always held desperation at bay. But panic was what he began to feel now, welling up inside him. For the first time since quitting his job at Johnson & Johnson, he gave serious thought to brushing up his résumé and interviewing for a job at a corporation. The thought sickened him, but he knew that soon he would have no other choice.

He cast about for something, anything, that could prevent the dream from crashing. One idea occurred to him: the sieves. We'll just have to market the sieves as a separate product, he thought.

Articles in the farm press about Mark's new sieves had generated a good deal of interest. Ralph was getting letters and calls every day from farmers who wondered if they could retrofit the self-leveling system in their combines. Although the sieves could be adapted to any machine, the cousins had resisted the idea of selling them as a stand-alone item. Mark, especially, opposed this strategy. He saw the self-leveling system as an important component in his supermachine. It was one of the features that would make the ultimate Bi-Rotor a revolutionary departure from anything else on the market. Selling the sieves by themselves would steal some of the thunder from his dream machine.

Ralph agreed, but argued that taking the sieves to market now appeared to be the only thing that could save them from failure. The sieves were the life raft that would keep them afloat. Mark was finally persuaded. His main task that fall and winter would be to refine the sieve system. They might begin manufacturing it, on a small scale, as early as the following spring.

Inquiries about the sieves were all going to Ralph's old address and phone number in Fort Worth. He didn't live there anymore, but he persuaded the new tenant of his apartment to let him keep an answering machine in a closet, and to forward his mail to the Kincaid shop in Haven. He called the machine often to check for messages. When he ran out of fund-raising leads to follow, he kept himself busy by responding to sieve inquiries.

One day the answering machine played back an unusual message. The caller was a man who said he had recently read about Mark and Ralph in a magazine called *Farm Industry News.* The man assumed that the cousins had already entered into an exclusive relationship with a major farm-equipment company. (He was so sure, in fact, that he almost didn't bother to call.) But on the off chance that he was wrong, he gave his name and phone number and asked them to call him back.

The man worked for a big company—bigger than either Deere or Case-IH. Ralph had never considered this company as a candidate for his sales pitch, though, because it neither made nor sold combines. "You'll never guess who just called," Ralph told me. "It was a guy from Caterpillar. They want to talk."

CATERPILLAR HAD LONG been the world's largest manufacturer of earthmoving equipment—bulldozers, excavators, loaders, graders, scrapers. By a process of synecdoche well known to the makers of Xerox machines and Kleenex tissues, the name Caterpillar had come to stand for any tracked machine that moved dirt, regardless of who made it. The trademarks Cat and Caterpillar brought to mind heavy construction, road building, mining, tree felling and clearing. The name did not, to most people, suggest agriculture or farming. But Caterpillar was moving, as slowly and cautiously as a Cat on the side of a cliff, toward diversifying into agricultural equipment. For decades John Deere, Case, and International Harvester had been raiding Cat's turf by adding construction equipment to their product lines. Now Cat was turning the tables.

Caterpillar's first venture into farm territory, introduced in 1987, was a tractor called the Ag Challenger. Challengers were big, beefy machines that competed against the largest class of farm tractors—articulated four-wheel-drive machines, 200-plus-horsepower land hogs. Unlike these tractors, however, the Challenger had no tires; it had tracks. Its tracks were similar to those on a bulldozer except that they were made of thick, steel-belted rubber, with rubber lugs projecting from them in a chevron pattern. The main selling point of these tracks was that they distributed the weight of the machine over a much greater surface area than tires did. It was like the difference between a snowshoe and a stiletto heel—one distributed weight, the other concentrated it. With tracks you got better traction, you didn't sink as easily in mud, and you caused the soil to be less compacted than you would with a rubber-tired machine.

The point about soil compaction was especially crucial. There is mounting evidence that compaction, caused by repeated trips across fields with machines weighing anywhere from ten to forty tons, has inflicted and continues to inflict serious damage on the nation's farmable topsoil. Compaction destroys the soil's natural porosity, creating, instead, a dense stratigraphy of hardpan. Healthy soil absorbs water like a sponge; compacted soil quickly becomes sodden, then repels excess moisture as runoff. Increasing agricultural runoff aggravates several problems, from soil erosion and flooding to elevated levels of pesticides in rivers, ponds, and lakes. Not surprisingly, soil compaction also reduces crop yields. Researchers at Purdue University recently found that crop yields on severely compacted soils were reduced by more than 50 percent.

Farmers who observed the effects of compaction on their land were becoming more and more interested in machines that were mounted on tracks instead of tires. Magazines like *Farm Journal* and *Successful Farming* carried feature articles about farmers who spent their winters converting entire fleets of equipment—tractors, combines, sprayers, grain wagons, slurry spreaders, the works—from tires to tracks.

Caterpillar encouraged this trend in any way it could. Its advertisements for Challenger tractors were text-heavy broadsides on the problem of soil compaction and the corresponding benefits of "running on Cat tracks." The message seemed to be getting through: The Cat Challenger muscled its way into a crowded market and was soon commanding a respectable share of tractor sales in the heavyweight class.

Caterpillar was still a bit player in the U.S. farm-machine industry, but the company's agricultural unit hoped to change that. Its small cadre of managers wanted to expand Cat's product line to include a broader range of tractor models. There was even talk of a Caterpillar combine. The company had participated in experiments to fit its rubber tracks on combines made by other companies, such as John Deere and Claas (a major European manufacturer based in Germany), but the resulting hybrid machines were too expensive to be competitive.

The tracks added something like $30,000 to the price of a machine that was already beyond the reach of many farmers. Somehow, Cat's agriculture managers concluded, they needed to find a simpler, less expensive combine to match with their track system. Nothing on the market would do. And designing a new machine from scratch wasn't practical for a company whose in-house staff of combine engineers numbered exactly zero.

About the time these ideas were churning around in Peoria, where Caterpillar has its headquarters, *Farm Industry News* ran its article about Mark and Ralph, along with pictures of Whitey. Cat's whiskers twitched; its ears pricked up. It sniffed cautiously. The phone rang in Ralph's closet. Soon he was deep in discussions with members of the company's agricultural research department.

Ralph felt an immediate rapport with his contacts at Caterpillar. They dealt with him frankly and as equals, projecting none of the calculated reserve, interpretable as snobbishness, that Ralph had picked up from representatives of other large corporations. "They aren't playing big cheese just to play big cheese," he said. Much of this deferential treatment, he realized, was a consequence of Cat's having very little expertise in combines. In other contexts Mark was just a farmer-inventor—a kind of idiot savant. But at Cat he was the acknowledged expert. He received the kind of respect and deference that were his due. If his innovations had concerned a bulldozer, however, Cat's attitude might have been different.

Caterpillar was a global company with more than 50,000 employees and $10 billion in annual sales and revenue. It was, however, in the midst of hard times. Demand for construction equipment was down because of a worldwide recession. In 1992 Cat was on its way to racking up losses of more than $190 million (its losses amounted to $2.4 billion if the effects of new accounting standards were included); the year before, it lost $404 million. It was deep in austerity mode, streamlining its factories, flattening its management structure, trimming its payroll. There wasn't a great deal of loose change lying around to spend on a new combine. Ralph's contacts in the company's agriculture unit told him that their departmental research budget for the year had been spent. They were definitely interested in the Bi-Rotor combine, they said, but they could not act on their interest until corporate budgets were renewed next year.

By this time it was October 1992. "Next year" was three months away. Ralph wasn't sure that he and Mark could survive that long, dangling, as they were, at the end of their financial rope. Maybe Cat couldn't put up any money now, but someone else might if they knew that a company of Caterpillar's stature was in the wings. He needed to extract some kind of bankable commitment from Cat before the year was out, so he proposed a meeting. "It's not just a new threshing system that we have to offer," Ralph told the people at Cat. "We've got some things that aren't patented yet that we think you'd be interested in. They could shape the direction of your agricultural division for the next fifteen years. If you'll sign a nondisclosure agreement, we'll show you everything."

Scott Larson drafted a two-page agreement in consultation with Caterpillar's legal department. Both parties signed the document, then set a meeting date.

The chosen venue was a room at the Holiday Inn in Great Bend, Kansas. Ralph prepared his flashiest performance to date, more elaborate even than his presentation to Rod McGuinn. He began with some slides. "Guys," he said in the darkened conference room, "we don't want a Bi-Rotor John Deere." He flashed a slide showing a red circle and slash mark circumscribing a big green combine—a kind of "Deerebusters" symbol. He let the Cat delegation absorb the implications of this for a moment. The message he intended to send was that a Deere Bi-Rotor was indeed a possibility if Cat did not act, and soon. Then he said, "We would much rather have *this.*" He clicked the slide changer and the image of an unfamiliar machine appeared. Mounted on crawler tracks, it looked rather like a top-heavy lunar transport vehicle. The machine was yellow with black trim—Caterpillar colors. On its side was a new logo that Ralph had hastily commissioned from Troy Robinson, a commercial artist he knew. It consisted of two intertwining arrows with the words "CAT BI-ROTOR" emblazoned across them. His audience grunted approvingly.

The next slide was a montage of photographs taken in the early 1900s and earlier. The photos showed giant wooden combines being pulled across vast fields by steam-powered crawler tractors. A headline on the slide said, CAT'S BEGINNINGS. "These were some of the first Caterpillars," Ralph said. "They're combining wheat in California. As you know, the company made combines for years. But you sold your combine division to John Deere in the nineteen-thirties."

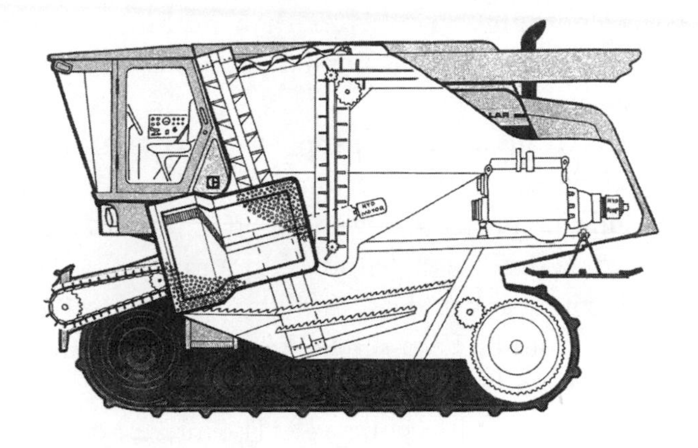

"It sounds like hamburger": A drawing from Ralph's presentation of the "ground-up" prototype. (Mark Underwood and Ralph Lagergren)

"And we've hated every minute of it since then," said a member of the Cat delegation.

"Well, it's time for you to get back to your roots," Ralph said. Then he flipped the room lights back on. He felt the meeting was off to a good start.

There was one skeptic in the Cat group, an engineer named Jeff. He kept asking technical questions about things like perforation diameters and centripetal coefficients. Mark answered the questions as well as he could, although Jeff didn't seem entirely satisfied. After more discussion, the group went outside and drove two and a half hours to a field where Whitey stood waiting. In spite of a heavy drizzle, Jeff inspected the combine with particular care, internally and externally. Then he sat in the cab as Mark ran several bushels of wet milo through Whitey's gut. Jeff was descending the ladder after this demonstration when another member of the group asked him, "So, how *does* this thing work?"

"I don't know exactly how it works," Jeff said, a note of exasperation in his voice. "It works! Let's just get her built."

After the Cat delegation left, Ralph felt a new surge of hope. "This is it," he said. "This is the light at the end of the tunnel." His sense of relief was qualified, though, because he knew there was still plenty of tunnel left. However ardent the new suitor seemed, Ralph had no concrete evidence that the relationship would go anywhere. He had learned the hard way not to bank on sweet nothings whispered in his ear. But sweet nothings were better than nothing at all, especially when they came from one of the fifty largest industrial corporations in the United States.

Best of all, he really liked the people he had met from Cat. "These guys were at the meeting to listen and learn," he said. "There was no superior attitude, no preaching. We hit it off. If John Deere came up to us tomorrow and offered us a definite deal, I'm not sure I'd take it.

"I'm not burning any bridges, though," he added. "I'd like to kick John Deere's butt, but I'll be kissing it if nothing else comes through."

As soon as he could, he told Bernard Taylor about the meeting. Taylor was glad to hear it had gone well. The interest of a big company like Caterpillar suggested new possibilities to Taylor. His idea was that Caterpillar would eventually subcontract the fabrication and assembly of Bi-Rotor combines to independent plants in Kansas. Cat would own exclusive marketing rights to the machine, and a small group of Kansas companies, with Taylor as their representative, would retain manufacturing rights. It would be like the old days, after the war, when American industry was in its glory and Taylor's plants made airplane parts for Boeing.

"This is fun," Taylor said, his eyes twinkling. "This is the kind of thing I live for. I want to see this deal go through if it's the last thing I do."

WAR ON WEEDS

Shall I not rejoice also at the abundance of the weeds, whose seeds are the granary of the birds?

—Henry David Thoreau, *Walden* (1854)

Sure, Henry, rejoice. And starve.

—Michael Pollan, replying to Thoreau in *Second Nature* (1991)

The ancient curse put on "the-man-with-the-hoe" lifted with the discovery of 2,4-D.

—John Schlebecker, *Whereby We Thrive* (1975)

Delbert Underwood, Mark's father, was forty years old—Mark's current age—when World War II ended. Delbert was one of the young lions of Jewell County, Kansas, always one of the first farmers to adopt new machines and methods. In the spring of 1948 he read a magazine article about a new weed killer called 2,4-D. The interesting thing was that the chemical didn't kill wheat or lawn grass, but it got rid of broad-leaved weeds like wild buckwheat, dandelions, and thistles.

The timing of this news could not have been better for Delbert Underwood. Just six months earlier, in the fall, he had planted about 200 acres of winter wheat. The autumn had been dry, and the wheat came up in thin patches. By spring the anemic wheat was heavily infested with ragweed, known locally as fireweed because it burns rapidly when dry. Delbert's fireweed problem got so bad that he thought he might have to plow up the wheat. But the magazine article about 2,4-D made him decide to try this newfangled chemical as a last resort.

No one else in Jewell County had used 2,4-D or any other synthetic herbicide before. Nobody knew how to apply such chemicals, and none of the farm-implement dealers in the area carried the equipment needed to do it. After a brief search, Delbert bought a sprayer from a company in Walthill, Nebraska, 120 miles northeast of Burr Oak. Then he bought two thirty-gallon drums of 2,4-D amine solution.

Delbert Underwood, who is now eighty-six years old, remembers his first experience with 2,4-D in detail. "It was the tenth of May," he recalled recently, "just before the wheat jointed. Always before that time, when you had thin wheat you had to sow oats or barley to try to crowd out the weeds. Or you had to destroy the crop by plowing it under. I wanted to avoid that, of course. So I got that sprayer and put it on an old Farmall H tractor. I dosed her pretty good—even shot a few patches twice. A few weeks later, I mean to tell you, those weeds were dead. 2,4-D saved my wheat crop that year.

"Of course," he added, "the old-timers frowned on it. They thought you'd ruin the ground for good if you sprayed that stuff. But 2,4-D wasn't like that. It didn't poison the ground, and it was amazingly effective. A lot of other farmers in the area started using it after I did, when they saw how it worked. It seemed like a miracle."

My grandfather, Elvin Canine, was one of the old-timers who frowned on "miracles" of that sort. Switching from horses to a tractor, and from open-pollinated seed corn to hybrids, was enough change for him. The ledger books in which he and my grandmother recorded their expenses contain no entries for 2,4-D or any other weed killer. If you were an old-timer who farmed on a small scale, as my grandfather did, then you could stay on top of the weeds well enough with a cultivator and a hoe.

But if you were a young buck with 200 acres of wheat and a similar acreage of row crops to worry about, weeds were a bigger problem. Hoeing was out of the question—the scale of the implement was all wrong. The proverbial "man with the hoe" was a perfect match for the man with a scythe, but no match whatsoever for the man (or woman) with a combine. Delbert Underwood owned a combine, having bought a John Deere model 5-A in 1934. By the beginning of World War II, he also owned two tractors and an assortment of tilling and planting implements. With these machines he could plow, plant, and harvest more than twice as much land as his father could at his age.

Delbert could have farmed even more land if it weren't for the weeds. Weeds could ruin a whole wheat crop as surely as a plague of locusts. Pests of all kinds, from weeds to insects to fungi, had become drags on the entire system of farming technologies, impediments to the further industrialization of agriculture.

The war itself helped to change that. 2,4-D was developed during World War II as a chemical weapon. It and a number of other compounds—the first man-made organic weed killers—were the products of an international arms race in chemical warfare that began thirty years earlier, during World War I, with the poison cloud of Ypres.

On Thursday, April 22, 1915, German mortar shells rained down on the city of Ypres, Belgium, for the third day in a row. Ypres was in the center of a skull-shaped salient in the Western front that extended deeply into German-held territory. After one major battle and months

Bull thistle.

of inconclusive trench warfare, the Germans decided it was time to eliminate the Allied salient at Ypres once and for all. Deafening salvos fell on the city every twenty minutes, turning the once-bustling Belgian gateway to the English Channel into a moonscape of craters and rubble.

Around midday, the shelling let up. It was a fine spring afternoon, sunny and warm. Allied troops, who were defending Ypres from a line

Buttonweed.

of trenches that nearly encircled the city to the north and east, could hear larks and thrushes singing. At five o'clock, a signal balloon carrying three flares rose above the German line. Allied soldiers thought this probably meant a new mortar attack was about to begin; they hunkered down in their trenches in anticipation of incoming shells. What came instead was a strange, soft hissing sound. Two greenish-yellow plumes appeared over no-man's-land. The plumes merged into a single cloud, which hugged the ground as it floated downwind, toward the shallow trenches where part of the Forty-fifth French Algerian Division was stationed. Soon the cloud dropped into the Allied trenches and spread. For five miles along the line, soldiers gasped, choked, and screamed as their lungs burned. Germany had just mounted the first major gas attack in military history.

The cloud of Ypres consisted of chlorine gas. Within minutes it had killed as many as 5,000 soldiers and incapacitated another 10,000. "I am not pleased with the idea of poisoning men," wrote one German eyewitness. "Of course, the entire world will rage about it at first and then imitate us." He was right. The British used chlorine gas for the first time a few months after the Ypres attack. But by then the Germans had graduated to phosgene, deadly at half the concentration of chlorine, so the Allies started using phosgene, too.

Soon all troops, Axis and Allied, carried protective masks and learned "gas discipline" in basic training. These defensive measures were effective, and blunted the strategic effectiveness of gas until, in 1917, the Germans unleashed another new poison weapon: dichlorodiethyl sulfide, better known as mustard gas.

Masks offered little protection against mustard gas. It seeped through clothing and blistered skin. By 1918 both sides were using mustard gas in large quantities. The U.S. War Department formed a Chemical Warfare Service and built a chemical plant at Edgewood Arsenal, near Baltimore. Soon after it opened, the Edgewood plant was pumping out toxic agents at the rate of 675 tons per week.

Public reaction against chemical warfare and its implied corollary, germ warfare, was immediate and strong. After the war ended, the voices of protest grew even louder. Postwar advances in aviation made aerial bombing an increasingly likely aspect of future armed conflicts. Aerial bombs, like mortar shells, could be filled with poisonous gas or, even more horrible to contemplate, agents of plague and pestilence. Unlike mortars, airplanes could penetrate far beyond military fronts, carrying deadly toxins and pathogens deep into civilian territory.

Gas and germ weapons were a matter of urgent debate at an arms-reduction conference held in Geneva in 1925. The conference produced the Geneva Protocol on Gas Warfare, which banned the military use of poison gases or agents of disease. Nonetheless, by the 1930s American intelligence sources had discovered that Germany, Japan, and Russia were all engaged in research on chemical and biological weapons. Diplomacy had apparently failed to put the malevolent genie uncorked at Ypres back into its bottle.

Germany had the world's largest and most advanced chemical industry and was especially successful in its efforts to develop new deadly agents. Working for the large German chemical cartel I. G. Farben, a researcher named Dr. Gerhard Schrader synthesized the first nerve gas in 1936. He discovered it by accident. Schrader was testing various organic phosphorus compounds for insecticidal properties when he found one that was exceptionally potent. In tests on mammals, the chemical caused paralysis of the nervous system, which led rapidly to respiratory failure and death. Schrader reported his discovery to the Reich's War Ministry, which was very interested. Schrader's chemical, the War Ministry discovered, incapacitated humans at far lower concentrations than mustard gas did. The new compound was called tabun. In 1938 chemists at I. G. Farben synthesized a second and even deadlier nerve gas called sarin.

Shadowy reports that Germany was producing new chemical and, perhaps, biological weapons prompted rival nations to act in kind. In 1942 President Franklin D. Roosevelt approved the creation of a civilian agency that would supervise research and development on biological warfare. To avoid alarming the public, the germ-warfare board was set up as an obscure agency-within-an-agency and given the nondescriptive name "War Research Service." George Merck, prominent head of the large American chemical and pharmaceutical firm Merck & Company, was put in charge.

Merck and his colleagues in the War Research Service soon informed the government that research on biological warfare would have to be conducted on an immense scale if it was to produce results in time to be of any use. No facility yet existed where a project of the scope and magnitude Merck envisioned could be undertaken. A large new center of research on biological warfare would have to be built. The site chosen for the facility was Detrick Field, a National Guard airfield just outside Frederick, Maryland. Construction proceeded quickly. By November 1943, the newly renamed Camp Detrick was pursuing its mission.

Frederick was a quiet town nestled in the eastern foothills of the Catoctin Mountains, a range of the Alleghenies. The town seemed rural and remote, though it was only forty-five miles northwest of Washington, D.C. The residents of Frederick could tell that something big was taking place on the military base at the edge of town, but they could only guess what it might be. Seeing smokestacks rising within the fenced compound, some townspeople thought it must be a chemical-munitions plant; others speculated darkly that it was an extermination center for Axis prisoners. Most Frederick residents, though, preferred to show no curiosity whatsoever about the base. All together, some 4,000 people worked there, including a few hundred civilians. The base's purpose remained one of the best-kept secrets of the war.

Between 1943 and 1945, Camp Detrick was the headquarters of the largest program of scientific research outside the Manhattan Project. The program at Detrick, in fact, competed with the Manhattan Project for certain kinds of scientists. A few government and military officials speculated that a secret weapon to end the war might emerge from Detrick rather than Los Alamos.

Detrick's military mission has been described as "public health in reverse." Researchers at the base tried to isolate and reproduce agents of infectious diseases, then figure out how to introduce the pathogens so they would spread rapidly in targeted populations. Some of the diseases investigated for this purpose were yellow fever, tularemia (rabbit fever), brucellosis (undulant fever), botulism, glanders, paralytic shellfish poisoning, anthrax, and pneumonic plague (the deadlier, airborne, version of bubonic, or black, plague).

In December of 1943, the U.S. War Department received intelligence reports suggesting that Germany might soon resort to biological warfare. Prompted by this threat, Secretary Stimson issued orders to enlarge the scope of research at Camp Detrick. Some existing programs were expanded and new programs added. One of the new additions was called the Crops Division. Its mission was to develop chemical and biological agents that could damage the enemy's ability to produce food and fiber. Among the cinderblock labs and incinerator stacks at Detrick, greenhouses bloomed.

The Crops Division had been the idea of Ezra J. Kraus, chairman of the botany department at the University of Chicago. For nearly ten years, Kraus and several of his students had been working with a new class of chemicals known as plant-growth regulators. These substances were synthetic versions of natural plant hormones. They affected plants in all sorts of odd ways. Some compounds, if applied to the stem of a plant on the tip of a toothpick, would make the stem bend, as if toward the sun. Some would cause a plant cutting to sprout new roots or new stems; others delayed the ripening of fruit or caused the formation of seedless fruits. Researchers were discovering more growth regulators and their remarkable effects every day.

In supervising students' work with these chemicals, Kraus had noticed that some of them were so potent that plants often died from accidental overdoses. Whenever this happened during a student's experiment, the student would throw the dead plants away and try the experiment over again with a lower dose of growth regulator. Kraus was apparently the first person to see this death-inducing ability as something other than a nuisance. He began to wonder if the chemicals might be useful as herbicides.

For the most part he kept the idea to himself, mentioning it to only a few colleagues. Before publishing anything, he wanted to test as many growth-regulating chemicals as possible for their effectiveness as plant killers. He enlisted the help of John W. Mitchell, a former student who worked for the USDA's Bureau of Plant Industry in Beltsville, Maryland. Late in 1941 the two of them—Kraus in Chicago and Mitchell in Beltsville—began testing various synthetic plant-regulating chemicals to see which ones caused the desired effect on various plants and at the lowest dosages.

By the middle of 1943 they had identified a number of promising compounds. One in particular stood out—a phenoxy acetic acid whose full-dress title was 2,4-dichlorophenoxyacetic acid. In time the name would be shortened to 2,4-D.

IN THE LANGUAGE of chemistry, 2,4-D is an aromatic compound. Aromatic, in this sense, means not necessarily that the compound has a strong smell (some aromatics do, some don't), but that it is one of a large number of compounds that have a structure based on the benzene molecule, C_6H_6—six atoms of carbon and six of hydrogen, arranged in a ring (which is graphically depicted as a hexagon). Chemists sometimes call this basic structure the "aromatic ring."

An infinite variety of chemical structures can be built using the benzene ring as a foundation. Different atoms or side chains can be stuck to the ring in various positions and combinations to create everything from moth killers to heart medications. The difference between two substances with strikingly different properties may be a few seemingly subtle changes around the aromatic ring—the substitution of a chlorine atom for a methyl group, for example, or the movement of a substituent from one position to another.

The 2,4-D molecule is a fairly simple one. The numbers in its name refer to two chlorine atoms affixed to the benzene ring at corner numbers two and four. A side chain, attached at position one, consists of an oxygen atom and an acetic-acid group. (Acetic acid is the main constituent of vinegar; it is also found in plant juices and human blood.) That's it. A chemist could draw a diagram of it as fast as you can say dichlorophenoxyacetic acid.

Plant-growth regulators, including 2,4-D and some other phenoxyacetic acids, are chemical impostors. They mimic indoleacetic acid (IAA), a natural plant hormone that stimulates growth. When these molecules come into contact with the roots, stems, or leaves of a plant, they act like party crashers with well-faked invitations. The molecules bluff their way past the plant's outer waxy layer and, once inside, circulate throughout it. They move from organ to organ, room to room, mixing with the crowd of nutrients and natural biochemicals. When they reach areas where cell reproduction and growth are taking place, the intruders get involved. Because of their molecular shape and electron density, they are able to fit certain biochemical receptors like skeleton keys, turning off, or speeding up, some of the plant's biosynthetic pathways. The results depend on which pathways are affected and how aggressively (and in what numbers) the party crashers interact with the plant's normal development.

Some plants can digest, or break down, various synthetic growth regulators before the intruders have a chance to do much mischief. This is what most grassy plants do to 2,4-D. Wheat, corn, sugarcane, turf and pasture grasses, and other members of the *Gramineae* family shrug off 2,4-D the way Superman shrugs off bullets. This "selectivity" is what makes 2,4-D so useful as a herbicide. It kills dandelions but not bluegrass, ragweed but not wheat, young velvetleaf seedlings but not corn. This property of 2,4-D—its selectivity and remarkable strength as a herbicide—is what Kraus and Mitchell got an inkling of in their research. It was a powerful discovery. Now Kraus had to decide what to do with it.

Normally, he and Mitchell would have published their findings in a scholarly journal such as the *Botanical Gazette,* which was published at the University of Chicago and which Kraus himself edited occasion-

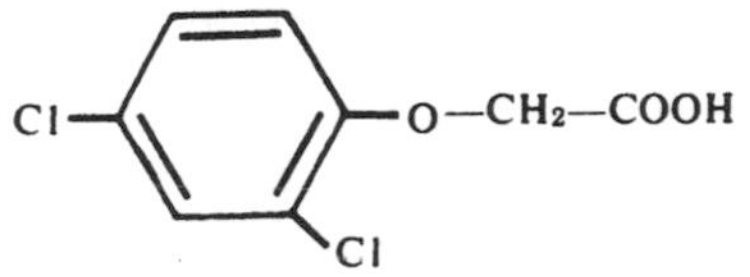

By mimicking a natural plant hormone, the 2,4-D molecule (whose structural formula is shown here) causes broad-leaved plants to grow themselves to death.

ally. But since the United States was at war, he reported his findings to the War Research Service, which ran Camp Detrick. Kraus thought the WRS might be interested, as he put it, in "the toxic properties of growth-regulating substances for the destruction of crops or the limitation of crop production."

This was the proposal that eventually resulted in the creation of the Crops Division at Camp Detrick. Kraus was appointed the division's top consultant in a massive scientific search for synthetic chemicals that could rain death on enemy crops. A team of at least fourteen scientists and technicians, with help from an unknown number of support staff, synthesized close to 1,100-growth-regulating compounds and tested them for herbicidal activity. All 1,100 chemicals were measured against the yardstick of 2,4-D for their herbicidal potency in at least three preliminary plant screens. The dragnet search identified dozens of compounds that were, in some tests, even more powerful than 2,4-D, which meant that they produced herbicidal activity at levels measured in parts per *million*. Nothing like this kind of potency had ever been observed before.

Herbicide research and development at Camp Detrick continued at full speed until Japan's surrender in September 1945. None of Kraus's chemicals were used in the war as a weapon, although it was close: American forces were poised to deploy a plant killer against the Japanese, but the results of the Manhattan Project made that unnecessary. "Only the rapid ending of the war," George Merck wrote in 1946, "prevented field trials in an active theater of synthetic agents that would, without injury to human or animal life, affect the growing crops and make them useless."

The work at Camp Detrick served as a springboard for the rapid development of the herbicide industry after the war. It also provided the technical foundation for the use of herbicidal weapons twenty years later. Then, in Vietnam, Kraus's idea blossomed into a bouquet of colorful code names, such as Agent Purple, Agent White, Agent Pink, Agent Green, and Agent Orange.

Even before World War II ended, word of the new chemicals and their herbicidal power began to spread. In August of 1944, two USDA scientists published an article in the prominent journal *Science* about using 2,4-D to kill weeds. They reported spraying 2,4-D on some orchard stock that had been infested with bindweed. Within ten days, the bindweed was dead.

If the two men, Charles L. Hamner and H. B. Tukey, were hoping to attract farmers' attention, they did well to focus on bindweed. Field bindweed was, and still is, one of the most common and obnoxious of farm pests. It's a perennial with deep roots and underground stems. Above ground, its long, twining vines choke out other plants. If you chop off the vines, new ones grow back in their place, like heads on the terrible Hydra of Greek myth. If you chop up the underground stems, each stem fragment produces a whole new plant. Before World War II, entire farms sometimes had to be abandoned when bindweed took over. This botanical Hydra seemed invincible.

Now Hamner and Tukey announced the arrival of a chemical Hercules. The mythical hero slew the Hydra by recruiting a friend, who burned each severed neck stub with a torch after Hercules cut off the heads. Apparently 2,4-D did something similar to bindweed by interfering with the plant's ability to regenerate new stems and vines. It killed the plant down to its root and was, in the strict sense, a truly radical solution to the weed problem.

Probably few farmers saw Hamner and Tukey's article in *Science,* but the popular press picked up on the story, and it spread like bindweed through a cornfield. Soon farmers were clamoring for more information about the chemical. But hardly anyone, including the USDA's own scientists and county-extension agents, had ever heard of 2,4-D. Those who did know about it, including John Mitchell and others at

Bindweed.

the Bureau of Plant Industry who had collaborated with Kraus, could not yet discuss 2,4-D, since their work at Camp Detrick bound them to rules of wartime secrecy. The situation was made even more chaotic because no official agency had established the safety and effectiveness of 2,4-D as a commercial product. No one even knew for sure whether it was a human toxin.

Government regulation of pesticides for their safety was practically nonexistent in 1944. The intent of the major laws pertaining to pesticides was not to protect public safety but to stop unscrupulous merchants from adulterating or mislabeling such old, inorganic pesticides as Paris green, lead arsenate, and bordeaux mixture. When new chemicals of unprecedented biological potency, like DDT and 2,4-D, began appearing during and after World War II, there was no real mechanism in place for testing and approving their safety. The first "tolerances" for protecting the public against unsafe chemical residues in foods weren't established until the 1950s. By then, however, hundreds of synthetic organic pesticides were already on the market. Scientists, whether inside or outside government, knew little about the effects of these products on human health or the environment. Pesticide regulation was a hopeless game of catch-up ball from the very beginning.

The scientists who first discovered 2,4-D and its powers were unanimous in believing that the chemical was totally benign, except for its intended plant victims. John Mitchell, for one, had been spraying the chemical in outdoor field tests with what, by today's standards, seems like reckless abandon. In August 1944—the same month in which Hamner and Tukey's bindweed article appeared—Mitchell and two colleagues sprayed 2,4-D on a clover-infested golf fairway at the Chevy Chase Club near Washington. The same trio also sprayed a large patch of dandelions on the Mall in Washington, D.C., in front of the Museum of Natural History. Nothing bad happened, apparently, except to the dandelions, which shriveled and died. The researchers acted on their intuitive hunch that 2,4-D was quite safe where people were concerned. Ezra Kraus offered the most dramatic proof of this: At a weed-control conference in 1945, he announced that he had eaten half a gram of 2,4-D every day for three weeks, and, he declared, he had never felt better.

Mitchell and his Beltsville colleagues were plant physiologists, not chemists, so they couldn't easily synthesize 2,4-D themselves. They found a convenient source at the American Chemical Paint Company in Ambler, Pennsylvania. Franklin D. Jones, a research chemist there, had been working with growth regulators for several years. By following the patent and technical literature, Jones, too, had zeroed in on 2,4-D. He was making it in quantity to be tested in paints and other formulations. In 1944 Mitchell and some Beltsville colleagues bought a pound of pure 2,4-D crystals from Jones's company for $12.50. Shortly after that, Jones filed for a "use" patent on 2,4-D as a herbicide. (John Lontz, a chemist at DuPont who later helped develop Teflon, had already patented 2,4-D as a plant-growth regulator, but Jones received his herbicide patent anyway.) Jones's timing was perfect. The bindweed article had stirred up interest and demand for a chemical whose herbicidal powers were still an official military secret. In 1945, the year Jones received his patent, his company introduced the world's first commercial systemic herbicide—a 2,4-D formula called Weedone.

The product was an immediate commercial success. Other companies soon jumped into the fray with 2,4-D formulations of their own, setting off an avalanche of patent suits that took years to settle. In 1945, the first year 2,4-D was sold to the public, total production of the chemical in the United States was 917,000 pounds. The next year production jumped to five and a half million pounds; in 1950 it was fourteen million pounds; and by the mid-sixties more than fifty million pounds of 2,4-D were made, purchased, and applied every year. Today farmers, lawn-care companies, golf-course superintendents, and home owners use more than sixty million pounds of 2,4-D annually, in 574 different products.

2,4-D IS THE oldest and one of the most widely used synthetic organic herbicides in the United States, and yet much remains unknown about it. Plant physiologists still don't know, for example, exactly how it kills plants. "The mechanism of action of 2,4-D has been studied more than for any other herbicide," says the *Herbicide Handbook* (1989), published by the Weed Science Society of America. "Investigation has shown that it causes abnormal growth response and affects respiration, food reserves, and cell division; but the primary mode of action has not been clearly established."

Nor has the Environmental Protection Agency (EPA) established to its satisfaction whether or not 2,4-D promotes cancer. A National Academy of Sciences report, *Veterans and Agent Orange,* published in 1993, assesses all the available evidence on the toxicity of the ingredients of Agent Orange, of which 2,4-D was one. (The other main ingredient of Agent Orange, 2,4,5-T, has been banned in the United States since the 1970s, because the process by which it is manufactured inadvertently produces a dioxin contaminant.) On the question of the carcinogenicity of 2,4-D, the report reviews all the relevant studies, then concludes with the equivocal language of risk analysis: "2,4-D thus presents a possible, but not probable, risk of cancer to humans."

There are no quick, clear-cut tests to determine whether a chemical causes cancer in humans. Cancer takes years to develop, and may result from long-term exposure to chemicals acting singly or in some synergetic combination. The link from effect to cause is extremely hard to trace and even harder to prove. One of the few certainties in the tangled skein of relationships between humans and pesticides is that farmers and farm workers are more prone to cancer than the general population. "Farmers have higher than normal rates of leukemia, multiple myeloma, non-Hodgkin's lymphoma and cancers of the brain, prostate, stomach, skin and lip," said Dr. Charles Lynch, an epidemiologist at the University of Iowa College of Medicine, at a recent press conference. "Chronic diseases like asthma, neurologic and kidney disease also may be related to agricultural exposures."

One of the accidental virtues of 2,4-D is that soil microbes break it down fairly quickly into nontoxic degradation products. Many other agricultural chemicals lack this virtue. The more long-lived chemicals stick around for months or years, remaining more or less intact as they move through the food chain or the soil. They wind up in places where they shouldn't be—in human milk, for example, where residues of DDT can still be found twenty years after the chemical was banned in the United States; or in remote wilderness lakes, where they are transported by raindrops; or in drinking water, where they wind up after leaching from the topsoil. One recent study reported residues of thirty-nine different pesticides and their degradation products in the groundwater of thirty-four states and Canadian provinces. Aldicarb, the most acutely toxic pesticide registered by the EPA, has turned up in well water in twenty-four states from California to Maine. Atrazine, the most frequently detected synthetic chemical in groundwater studies, is a suspected human carcinogen. Nitrates from the overapplication of synthetic fertilizer regularly turn up in municipal water supplies throughout Iowa and other farm states at levels exceeding the EPA maximum. Conventional water-treatment plants do not remove nitrates or pesticide residues from drinking water.

PARTLY BECAUSE OF legal and public pressure, and partly out of enlightened self-interest, pesticide manufacturers are slowly phasing out some of their older, more environmentally harmful products. To replace them, the companies are spending hundreds of millions of dollars to develop a new generation of "environmentally friendly" chemicals. Such pesticides are effective at extremely low rates, dissipate quickly in soil, don't show up in drinking water, and are no more toxic to humans than grain alcohol or milk of magnesia. In short, they are as removed from the first generation of pesticides as 2,4-D was from old-fashioned mixtures of lead and arsenic.

The first members of this next generation of weed killers have already arrived. They are a group of chemicals called sulfonylureas that were discovered by a DuPont chemist named George Levitt. The chemicals themselves represented a huge advance over anything previously known, but the process of their discovery was much like the massive scouting effort that had taken place at Camp Detrick thirty-five years earlier. The main difference was the setting: While the Detrick researchers had labored in stark, confined cinder-block quarters on a military base, Levitt worked within the ivy-covered walls of a state-of-the-art lab at DuPont's Experimental Station, one of the largest industrial-research centers in the world.

Levitt made his first sulfonylurea in 1957 shortly after he joined DuPont's agricultural-products division as a research chemist. His job, and that of many other chemists who worked at the Experimental Station, was to brew as many novel compounds as possible. These compounds then went to the screening department, where they were tested for various kinds of biological activity. There were screens for herbicidal, fungicidal, insecticidal, and plant growth-regulant effects. A compound that caused even a slight "wiggle" in one of the screens constituted a chemical lead. When a chemist found a lead compound, then he or she (or sometimes a different chemist) worked with it to see if, by a little molecular tweaking, its biological activity could be amplified.

Some corporate-research chemists, like Levitt, spent their entire careers without producing a lead that went anywhere. Their job was like shooting in the dark—there was a good deal of luck involved. Many of Levitt's colleagues figured it was a numbers game: The more compounds you synthesized, the better your chances of hitting a target. But Levitt was slower and more deliberate than many of the others. He might make around 150 new compounds a year, while some friends of his turned in twice that many. He read the chemical literature a lot, which helped him choose targets before shooting. He knew, for example, that a popular drug for diabetics, called Orinase, had an interesting molecular structure with millions of possible variants. The common denominator was a side chain beginning with sulfur that connected two benzene molecules like a bridge. The molecular formula looked like a barbell: ring A, bridge, ring B. The bridge gave this group of chemicals the generic name sulfonylureas.

If Orinase was biologically active (which obviously it was, or it wouldn't be a drug), then maybe you could produce some wiggles if you played around with this barbell. Levitt decided to try that for a while to see if anything would happen.

Nothing did happen. None of the Orinase analogs he made caused even the slightest wiggle in the screens. So Levitt filed away the chemicals he had made in storage and hit the chemical abstracts again. Soon he had settled on another promising direction in which to aim his sights.

Seventeen years later, in 1974, another DuPont scientist—an entomologist named Cy Sharp—sifted through several older compounds on file at the Experimental Station. He was looking for chemicals that killed mite eggs. His search pulled up Levitt's sulfonylureas, which he tried in his mite-egg screen. One of the compounds, which had the code name E-707, produced a weak wiggle. Sharp told Levitt about the wiggle. Levitt decided to play with E-707 to see if he could upgrade the activity. He made several related compounds, but none of them worked any better than E-707 in Sharp's mite-egg test. He also sent his new batch of sulfonylureas through the normal battery of screens. Word came back that one compound created a wobble in one screen: It retarded plant growth. The effect was weak. At a dosage rate of two kilograms per hectare (roughly two pounds per acre), plants sprayed with the compound grew about half as much as the unsprayed controls.

"This struck me as a chemical lead," Levitt told me.

He described these events, which had taken place twenty years earlier, as he sat in my room at a Hilton hotel near Wilmington. Levitt had retired from DuPont in 1986, thirty years from the day he started. Now in his late sixties, he was dressed casually in khaki pants, a dark-blue turtleneck shirt, and a teal jacket that set off his blue-gray eyes. His face was full and craggy, spanned by a pair of metal-rimmed glasses and topped with sandy gray hair.

"The growth-retardant effect was so minimal that I didn't bother to report it to management in my research review," he continued. "But I thought it might be important, so I went to Ray Luckenbaugh, the leader of my group. He was a wonderful cheerleader, always telling us our ideas were great, urging us on. So I felt encouraged to pursue the lead and see where it might take me."

With a logical program in mind, Levitt and his lab partners synthesized hundreds of compounds. "Once you know what you want to make, you just play around a little bit with different starting materials and intermediates," he said. "You start with *ortho*-cresol, oxidize that with chlorine, which changes it to sulfonyl chloride, which then, treated with ammonia, gives the sulfonamide. Then we treat this with phosgene to get the sulfanylisocyanate, which is an extremely active compound. You make a bunch of this stuff and go on from there. It's a cookbook sort of thing."

Each week Levitt's group sent a batch of new compounds down for screening. Any of the compounds could have been anything—a diabetes drug, an insecticide, a plant-growth regulator, a weed killer, a rat killer—no one could predict. For every drug or pesticide that makes it onto the market, many thousands of compounds were screened and rejected. During the 1970s, the ratio of screened compounds to commercial products averaged about 7,500 to 1. During the next two decades that ratio grew steadily higher as chemical products were expected to meet more exacting standards of performance, precision, and safety. Today DuPont estimates that it screens more than 40,000 compounds for every product that makes it to the marketplace.

As Levitt synthesized compounds, he tagged some that he suspected were most likely to cause wiggles in the herbicidal screens. The tags were intended as signals for Dave Fitzgerald, the biologist who supervised DuPont's screening tests. Fitzgerald's main tool was a device that resembled a small automatic car wash, or a commercial dishwasher that trays pass through. For every compound to be screened, technicians prepared a tray containing several soil-filled peat pots. Every tray was the same. Some of the pots on the trays contained seedlings of various species; in other pots, seeds were freshly planted. The trays were placed on a conveyor belt that went through Dave's car wash—a fifteen-foot-long, Plexiglas-covered spraying chamber. Nozzles in the chamber were calibrated to spray chemical solutions at various prescribed rates, such as two kilograms per hectare. After each tray went through, the spraying system automatically rinsed itself. Then came another tray, which was sprayed with another compound, and so on. After the trays were treated, they were put inside a growth chamber or a greenhouse and monitored for signs of unusual activity—seeds that wouldn't germinate; seedlings that died, got sick, dropped their leaves, or otherwise behaved differently from the untreated control plants.

"Dave Fitzgerald was ordinarily a pretty serious guy—not very excitable," Levitt recalled. "But he started coming back with reports of increasing herbicidal activity. Every week he would come to us with a big hurrah. We were closing in on something."

Levitt would try tweaking his compounds in one direction, which might lead to reduced activity; then he moved instinctively in some other direction until Fitzgerald began reporting bigger wiggles. The trick, as Thomas Midgley, the internal-combustion pioneer, once remarked, was to change a wild-goose chase into a fox hunt. The process built on itself. Colder, warmer, colder, warmer, warmer—Levitt homed in on increasing activity like a self-guided missile. A pattern began to emerge, a predictable relationship between molecular structure and activity. Wiggle turned to jiggle, which turned to jump.

"Then one day in June 1975," Levitt said, "I got a call from Dave. He said, 'You've got to come down and look at this.' One of the compounds we'd sent him was so potent that minor residues in the spray system were injuring plants in subsequent tests with other compounds. It took him a while to figure out what was going on, and when he did he could hardly believe it. Dave didn't usually get excited, but he was excited."

The code number of the compound that produced all the excitement was R-4321. Levitt nicknamed it Countdown. By association, all the sulfonylureas Levitt's group tested after that were known as the "Countdown chemicals."

He had never felt more exhilarated by his work. "I just went from one high to the next," he remembered. "I felt that even if we didn't get a commercial product out of this, we had made a scientifically significant discovery."

The hurrahs from Dave Fitzgerald grew louder when Levitt began substituting methyl groups (CH_3) in various positions around ring B of the barbell. He had been playing along these lines when he came up with Countdown. As a skeleton key for gaining admission to some crucial biopathway in plants (they didn't know yet which pathways were affected), the Countdown molecule had something special going for it. It was as potently active as any herbicide on the market.

Still, it was a long way from making the grade as a commercial product. To be marketable, a compound not only had to be active, it had to be active in some economically useful way. Countdown, at the rates tested so far, showed little or no selectivity—it killed plants fairly indiscriminately. That restricted its usefulness to farmers. Also, by this time, any candidate for product status had to pass a tough battery of toxicity tests. And it had to degrade quickly in the environment, breaking down into chemical constituents that literally would not hurt a flea.

On most of these points, Countdown was a little rough around the edges. In itself it was an interesting scientific discovery, but it was not yet a million-dollar molecule. To make it that, Levitt and his lab partners would have to polish and refine it. They had moved from the scouting stage to the optimization stage. The task before them now was to synthesize and screen hundreds, perhaps thousands, of analogs of the Countdown compound. Levitt successfully lobbied for another chemist to help with the synthesis.

The direction their exploration should take, Levitt thought, was clear. They had found what seemed to be the optimum configuration for ring B of the barbell; now it was time to work on ring A. One by one, in systematic fashion, they substituted entire families of chemical groups in various positions around ring A. They tried tolyl groups, fluoro groups, chloro, trifluoromethyl and nitro groups. Once again, they began to observe relationships between structure and activity, and these observations informed the process. In the course of trying analogs with yet another group—esters this time—Levitt synthesized one that had the ester group CO_2CH_3 attached to a position adjacent to the sulfonylurea bridge. Once again, Dave Fitzgerald got excited. This was the most active compound yet. Fitzgerald lowered the application rate of the compound in order to find the bottom—the rate at which herbicidal activity diminished, or at which the compound stopped killing everything and showed some selectivity. But there didn't seem to be a bottom. No one had ever seen a molecule as deadly to plants as this one was. Levitt was on a high again.

"There was no precedent for the kind of activity we found, no precedent for the types of chemical structures that were involved here," Levitt told me.

I asked him, Why? What was going on at the molecular level to cause such activity?

"It's an electronic effect on that heterocyclic ring," he said. "There's a great abundance of 'surplus' electrons—a high electron density, similar to that of sulfa drugs—that's responsible for the activity."

Around the time Levitt made his crucial ester substitution, kicking the molecule's activity up to new heights, he was also quietly urging corporate managers to expand the small sulfonylurea group. Management, however, had other priorities, and decided instead to cut the group from six chemists down to three. Their reasoning was simple. The compounds Levitt and his partners were synthesizing required complex chemistry and would therefore be expensive to make. The retail cost of these chemicals per pound would have to be extremely high—so high that no one would buy them. Sulfonylureas, however interesting from a scientific point of view, could never be profitable products. It was time to cut bait before DuPont wasted any more expense on chasing this fascinating but commercially worthless quarry.

A crucial assumption behind the decision to dissolve the Countdown group was, it turned out, wildly off the mark. The assumption was that sulfonylureas would be applied at rates of somewhere between a quarter of a pound and one pound of active ingredient per acre—a low rate at the time. Most herbicides were applied in the range of a pound to a pound and a half per acre. The managers who projected the cost of sulfonylureas felt that they were giving the chemicals the benefit of the doubt by attributing an unusually high potency to them.

But the compounds turned out to be far more potent than anyone dared to think they could be. Levitt's best compounds, tests soon showed, were effective at a mere fraction of an *ounce* per acre. One compound, which Levitt synthesized in February 1976, killed weeds at rates of a few *hundredths* of a pound per acre (that compound later became Glean). When Dave Fitzgerald ran these chemicals through his car wash, he had to reduce his screening rate from the standard two kilograms per hectare to a mere whiff—0.05 kilograms or less per hectare. The difference in active dosage between Levitt's sulfonylureas and the most powerful existing herbicides was comparable to the difference between a gallon jug and a tablespoon.

Once the sulfonylureas' phenomenally low rates of application were plugged into the equation, the economic picture changed drastically. These chemicals might be expensive to make, but the cost to farmers per acre of weed control could still be quite competitive. Most herbicides were (and many still are) sold in bulky two-and-a-half-gallon jugs. The equivalent dosage of a sulfonylurea could be packaged in a container the size of an aspirin bottle. Less of the chemical would have to be made, shipped, stored, transported to the field, mixed, and applied. Importantly, less of the chemical would wind up dispersed in the environment. From the manufacturing plants that made sulfonylurea herbicides, there would be no effluent.

The potency of the chemicals comes, in part, from their very specific mode of action. When a small dose of one of them is sprayed on a plant in dilute suspension, a few droplets land on the leaves, stem, and soil. Only about 20 percent of the active ingredient that comes in contact with the plant eventually penetrates its waxy outer coating and gets inside. The sulfonylurea molecules make their way into the plant's plumbing, where they move up and down along with water and sugars. Before long, some of the molecules arrive at the meristems—places at shoot and root tips where the plant is generating new cells. Plant cells are like small balloons that inflate as they develop. When a sulfonylurea molecule arrives on the scene, it stops the balloons from being inflated. It does this by inhibiting the action of an enzyme, acetolactate synthase (ALS), which the plant needs in order to make three essential amino acids. It takes very little of the sulfonylurea to cause significant ALS inhibition. This has an immediate and devastating effect on cell division in the meristems. Suddenly—*screeech!*—the plant stops growing. It becomes a green skeleton. In time it succumbs to stress. As one DuPont scientist told me, "Weeds that aren't growing eventually just go kaput."

Some plants contain a mechanism that acts like Pac-Man and eats up sulfonylurea molecules before they cause the plant any harm. These plants metabolize the chemical into harmless by-products. They are resistant to the herbicide.

Animals don't have acetolactate synthase or the biochemical pathways in which it acts. They are therefore immune to the effects of ALS inhibitors like sulfonylurea herbicides. To animals, Levitt's compounds are roughly half as toxic as table salt.

From a low point in 1975, when Levitt's group was about to get the ax, the project suddenly grew into a major research initiative. Dozens of scientists were now being assigned at all stages, from synthesis to screening to toxicology and environmental testing. All told, DuPont poured about $35 million into the program during the next six years.

Companies like DuPont produce, on average, one commercial

product for every two synthesis chemists they employ. A chemist who originates one major product during his or her career is considered highly productive. Levitt's dogged efforts in his obscure backwater of chemistry eventually produced for DuPont fourteen commercial and advanced-candidate sulfonylurea herbicides. He himself synthesized four of them—Oust, Glean, Ally, and Harmony.

Before the potential of sulfonylureas became evident, Levitt was considered a so-so chemist—a genial man of persistence, but not brilliance, whose annual output of novel structures was below par. Now there is a building named after him on DuPont's agricultural-research campus in Newark, Delaware. The only other DuPont chemist to have a building named after him is Wallace Carothers, the inventor of nylon.

I asked Levitt if he feels like a superstar.

"No," he replied, smiling. "I just feel like a journeyman chemist who came out smelling like a rose."

We had been talking for almost three hours. Before our interview, Levitt had dropped off his son for a medical appointment.

"It's about time for me to pick up my son," he said. "But first I'd just like to say that I feel, with the sulfonylureas, that we have a class of chemicals that play an important role in agriculture, and that they're environmentally safe. I'd like to think that I've done something worthwhile, in spite of all those who are detractors of all chemicals, who are generalizing from specifics. If they took some time to look at sulfonylureas, maybe they would see some good.

"There's so much chemophobia out there," he added, standing up to go. "Such a fear that one molecule can make all the difference."

Some critics of synthetic pesticides would like to see all of them completely eliminated from agriculture. Even if everyone agreed that this is a desirable goal—rather like the complete elimination of nuclear weapons—in practice it would create some serious problems. The most daunting of these problems stems from one fact: The population of the world is growing at the annual rate of ninety-two million people, adding the equivalent of a new Mexico every year, mostly to Third World countries. It's not at all clear that, without some chemical herbicides, insecticides, fungicides, and fertilizers, the food needs of all those people could be met. Although human inventions have done little to eliminate the political causes of starvation from the world, technology has managed to drive the Malthusian wolf from the door—so far. But its howling can still be heard from just over the hill.

A second major problem with banning all agricultural chemicals, in the short run, is that it would lead to chaos on the land. Farmers would go out of business, farms would be overrun with pests, and crop production would decline precipitously. The fact is that chemicals embody a certain kind of bottled intelligence, and eliminating them would represent a net loss of intelligence from conventional farming systems. The farmer must make up for the loss, and then some, by reinjecting a different kind of intelligence into the system—a subtle intelligence that attempts not to overcome nature but to direct natural processes to human ends.

The more subtle kind of farming I am talking about goes by various names. Its purest form is organic farming, whose practitioners use no synthetic pesticides or fertilizers. It takes knowledge, conviction, and several years of gradual change to convert a conventional farm into an organic one. For many growers who live near large urban areas, the effort has paid off. There's a substantial niche market for organic produce in many cities. A Harris poll conducted for *Organic Gardening* magazine in 1989 found that 84 percent of the households surveyed would buy organically grown fruits and vegetables if they cost the same as regular supermarket produce. But that last "if" is a big one. Organic agriculture is not just a set of techniques; it has an attendant value system in which dollars and cents, while not ignored, are not the ultimate arbiter of worth. Most of us, however, do not take this alternative value system with us when we shop at the supermarket. Americans spend less of their incomes on food than anyone else in the world, and we like it that way. Since organic agriculture rejects the chemical-based approach of mainstream agriculture, it can rarely compete on the basis of price alone. Many supermarkets have tried carrying organic produce alongside their usual offerings, but most stores have eliminated this practice because the more expensive, and sometimes more blemished, organic produce doesn't sell.

A middle ground between organic farming and chemical-intensive agriculture goes by the name "integrated pest-management," or IPM. Farmers who practice this approach borrow some techniques from the organic school, such as using beneficial insects instead of insecticides, green manures and compost instead of synthetic fertilizers, and mechanical cultivation and cover crops instead of herbicides. But they do not shun chemicals entirely; indeed, they embrace much of the latest in chemical and mechanical technology to reduce the load of the most harmful (which are mostly older) chemicals in the environment.

IPM programs vary from region to region, crop to crop, and farm to farm, but a common thread that runs through all of them is the idea of an "economic threshold," below which pest populations are tolerated. Instead of following a fixed spraying schedule based on worst-case scenarios, which is what conventional farmers do, IPM practitioners monitor their fields with weed counts and insect traps. They use chemical sprays as a last resort when monitoring shows that a pest is multiplying above the economic threshold. Even then, sprayings are carefully timed according to pest life cycles and done at the stage (such as during larval hatch, when hungry caterpillars are easiest to kill, but before they have done much damage) when the chemical will be most effective, and at the lowest doses. Simple monitoring and more careful timing of sprays have allowed American cotton growers, for example, to cut their insecticide use in half since the late 1960s. IPM techniques not only reduce the load of pesticides released into the environment but also save farmers the cost of unnecessary pesticides.

The IPM concept has spawned a whole new range of high-tech farm machines. Instead of spraying entire fields with three or four herbicides, each one effective against different kinds of weeds (which is the common practice today), some farmers are already hiring companies with "smart applicators" to do their spraying for them. A smart sprayer is a vehicle that carries separate tanks containing several different herbicides. The nozzles and valves on the machine are controlled by a computer that contains a digital map of the field in its memory, with specific information about which kinds of weeds grow where. The smart vehicle keeps track of its position in the field by bouncing signals off satellites in space. It dispenses the proper amount of the proper herbicide only where it is needed.

High-tech sprayers like these represent the cutting edge of a trend toward adopting global-positioning satellite (GPS) technology in agriculture. Most applications are still in the experimental stage, though some GPS features are poised to hit the market within the next several years. Soon it may not be unusual to see combines equipped with satellite locators and computers that keep track of grain yields from each small sector (say, a square rod, or 272 square feet) of every field. This information could be put to good use in various ways during subsequent growing seasons. The idea is to break down a large field into small subunits and then implement the old socialist creed, "from each according to ability, to each according to need." High-producing areas of the field would receive more fertilizer, more seeds, and more crop-protection chemicals. Less-fertile sectors would receive less of every input, since high levels would be wasted on them. Advocates of this high-tech vision call it "site-specific farming," "prescription farming," or "farming by the yard."

One enthusiastic advocate of these ideas is Alan Van Nahmen, who would eventually join Mark and Ralph's team of Bi-Rotor builders. Van Nahmen has collected a thick sheaf of articles and papers on farm vehicles of the future, a subject that he finds fascinating. All signs, he believes, point to the development of single, multipurpose machines that will perform every major farming operation with the help of computers and satellites. In the spring, such a machine would be fitted out with a no-till planter or drill in front, and perhaps with a smart sprayer and fertilizer applicator in back. In the fall, the same vehicle, with an intimate computerized knowledge of its owner's fields, would gather the harvest, collecting a new set of yield and weed data along with the crops.

Today it's not unusual for a farmer with a tractor, various tillage tools, a sprayer, a cultivator, and a combine to make ten or eleven passes across each field in the course of a growing season. The multipurpose vehicle that Van Nahmen envisions, by contrast, would do everything in just two passes—one in the spring, one in the fall—thereby saving fuel, reducing soil compaction, and making better use of farmers' capital. This multipurpose machine, he thinks, will look very much like Mark's pared-down, superefficient combine. That's why Van Nahmen decided to join forces with Mark and Ralph. He sees the Bi-Rotor, in fact, as far more than just a combine. To him it's the Farmall of the twenty-first century.

DRESSING THE BRIDE

Follow a shadow, it still flies you;
Seem to fly it, it will pursue:
So court a mistress, she denies you;
Let her alone, she will court you.

—Ben Jonson, "Follow a Shadow" (1616)

By December of 1992 the cousins were, in Ralph's words, "hanging on by our fingernails." Their prototype sat in mothballs, and their checking-account balance hovered near zero in a persistent vegetative state. Caterpillar was by far their best hope, but the corporate machine moved at its own stately pace, oblivious to Mark's and Ralph's distress. The company continued to flirt and sniff with interest, but the sniffs had yet to escalate into a full-blown mating ritual.

Mark and Ralph did their best to foreshorten the courtship and move straight to nuptial arrangements. In particular, they pressed for a financial commitment from Caterpillar. The company's fiscal rules, however, could not be bent. Budgets would not be renewed and reallocated for months yet. But there was also the question of whether Mark's machine would perform as advertised. Caterpillar wasn't about to make a major commitment to Mark and Ralph until the company knew what it was getting. That couldn't happen until the prototype was built, but the prototype couldn't be built until the cousins had new financial backing. It was a classic catch-22, and the cousins' fingernails wore short as they awaited a resolution.

To complicate matters further, Mark and Ralph needed to find a new home for their project. Things simply hadn't worked out at Kincaid Equipment Manufacturing. For one thing, the cousins were paying Kincaid for parts fabrication by the hour, and this was proving too costly. Also, Mark needed more shop help, more technical support, and more space than the Haven company had to offer. If Mark and Ralph had learned anything over the preceding year, it was that designing and assembling a complete combine was more than a one-man job. They needed the resources of a larger machine shop. Their relations with the people at Kincaid remained cordial enough, but the strain was beginning to show. If the prototype was ever going to be finished, it would have to happen somewhere else.

But where? The ideal facility would be located in central Kansas, it would afford the combine project more shop space than Kincaid could spare, and its owners would want to assume a larger financial role in the partnership. After Ralph and Mark had exhausted their own short list of possibilities, Ralph called Bernard Taylor. Taylor said he knew of a place that might take them in. It was a company called Gordon-Piatt Energy Group, which made burner units for commercial and industrial boilers. Gordon-Piatt had done well in this niche, but like most other businesses, the burner trade had its ups and downs. Gordon-Piatt's managers were looking for ways to diversify their product base so they could keep the plant running two or three shifts a day instead of just one. To that end, Taylor had already helped Gordon-Piatt procure some aviation contracts. The company now had a nice little sideline making pilot and crew seats for Beechcraft and Cessna. Now Taylor steered Mark and Ralph to Gordon-Piatt's doorstep. He arranged for Ralph to pitch the combine to the president of the company, Jim Salomon.

Salomon was a compact man with black hair, a neat mustache, and so many gold rings on his fingers that they grated against one another when you shook his hand. He had worked his way up through the ranks at Gordon-Piatt from the warehouse floor and was widely admired for his skills as a businessman.

When Bernard told Salomon about the Bi-Rotor combine, he was interested. Agriculture dominated the economy of southeastern Kansas, where Gordon-Piatt was located, and an agricultural machine would bring the desired diversity to Gordon-Piatt's business. Also, Salomon respected Bernard Taylor's judgment. The old manufacturing man's high opinion of the combine venture carried weight. Soon after Ralph met with Salomon to pitch the machine, Salomon invited the cousins to move in under Gordon-Piatt's roof. The understanding among them was that Gordon-Piatt would provide shop space, raw materials, and fabrication labor for the combine prototype. In return, the company would secure manufacturing rights to a significant portion of the combine's parts—assuming the machine ever went into production.

The agreement was still verbal and informal when, one gray winter day, Mark and Ralph unceremoniously cleared out of Haven. They hauled their belongings—camping trailer, Whitey, unfinished prototype, salvaged International 915 carcass, tools, and several black garbage bags filled with dirty clothes—to their new base of operations. They told no one of their whereabouts except family and close friends.

Gordon-Piatt sits midway between Winfield and Arkansas City, Kansas, about ten miles north of the Oklahoma border. The plant is a sprawling mass of low buildings punctuated, in profile, by several tall boiler stacks and other steeples of industry. Out front, near the divided highway linking Winfield and Ark City (as Arkansas City is known locally), a neat one-story building faced with red brick houses the company's executive offices. The Gordon-Piatt complex occupies one corner of Strother Field, a former National Guard air base converted to a public airport and industrial park. Fields of milo and wheat surround the airport, stretching away to the horizon over a flat, flinty landscape.

When Mark and Ralph arrived at the Gordon-Piatt plant with their gear they were shown to an unmarked building at the north end of the Strother industrial area. Gordon-Piatt was in the process of acquiring this building from another company. The corrugated-steel structure, which was nearly empty, enclosed an area about the size of a football field to a height of forty feet. Its flat shed roof sloped almost imperceptibly. Though of middling size by industrial standards, the mustard-colored building was large enough to swallow the cousins' assorted fleet of vehicles, plus any amount of additional clutter they might generate, and still appear vacant. Mark scouted the vast concrete floor for the least conspicuous place to set up shop, like a pregnant cat deciding where to have her kittens.

Once the cousins had solved their housing crisis, they turned back to their financial one. This meant courting Caterpillar as vigorously as possible. Ralph could be a very charming and persistent suitor. He wooed prospective investors and corporate partners with the same ingenuity and fancy footwork that made him a popular dance partner. With Cat, he called as often as several times a week to see what was going on, often playing up some new aspect of Mark and his miracle machine.

Ralph's main contact in Peoria was a man whom I will call Bob Petersen, whose business card read: "Program Manager, Ag/Forest Products Research Department." He was an upper-midlevel manager whose placidly nerdy exterior belied a strong ambition to distinguish himself at Cat, where he had spent his entire twenty-five-year career as a mechanical engineer. To bring an innovative, successful new combine into Cat's product line, preferably under budget and ahead of schedule, would be a brilliant achievement, one that was bound to earn him prestige and advancement within the company. But the effort could also bomb. Petersen was not a combine expert, and he didn't want to venture too much time or money on Mark's Bi-Rotor design until he felt assured that it could live up to the cousins' optimistic claims.

Petersen wanted the prototype to be built and tested as soon as possible, so he responded to Ralph's full-court press with an assertive gesture of his own. It would still be some time, he told Ralph, before the company could send any cash relief for the cousins' checking account. Petersen could, however, promise something almost as good: He offered to send down some crucial hardware for the new Bi-Rotor prototype, including a Cat diesel engine, a new set of Cat tracks, and an undercarriage assembly on which to mount the tracks.

On hearing this offer, Mark and Ralph were all smiles. They felt the hot ecstasy of a love-smitten youth whose object of adoration consents, after weeks of wooing, to hold hands.

One condition of a commitment from Cat was that the new machine had to ride on tracks. Mark had, however, designed and built his prototype-in-progress to ride on four wheels, like other combines. He and Ralph realized that adapting the prototype to tracks would require changes of a major order. In fact, they determined that simply starting over would be easier and ultimately better from an engineering standpoint. That meant abandoning everything they had done at Kincaid during the preceding fourteen months. Painful as this prospect was, the decision to embrace it wasn't difficult. "Mark always wanted to put the machine on tracks," Ralph said, "but we didn't think we could afford it. Now we've got a free set of tracks. It's just one more thing that's fallen into place."

Nobody knew yet what the track system for their combine would look like. There was a big question mark lurking under the machine.

Indeed, the offer of tracks made the cousins a little nervous. Details of the undercarriage would have to be worked out before Mark could design what rode on top of it. And he had an inkling that collaborating with Cat engineers on the track system might not go entirely smoothly.

Mark, Ralph, and Bob Petersen agreed that they should meet to discuss the matter. Ralph suggested getting together on neutral territory, somewhere between Kansas and Illinois. "How about Des Moines?" he asked Petersen. One reason Ralph suggested Des Moines was that he knew he and Mark would have a place there to stay—my house.

"I'M MORE NERVOUS about this meeting than I was when we first met with Cat in Great Bend," Mark said, staring into a crackling fire. "We hit it off on our first date. Now we've got to learn how to dance without stepping on each other's toes."

Mark, Ralph, and I were sitting around a wood stove at my farm. It was a chilly night, eleven days before Christmas. Sipping beers, we talked about the meeting that would take place the next morning.

"We have a dilemma," Ralph said. "They've got this half-track system they want us to use. It's got tracks in front and a pair of wheels in the rear. The wheels are for steering. When you turn the steering wheel, the rear tires come down and lift the back end of the machine up. The combine sort of tippy-toes from side to side."

The half-track system was an expedient, designed to save money. Machines with full tracks, such as bulldozers and military tanks, steer by changing the relative speed or direction of the tracks. This is called differential steering, and it requires each track to have its own transmission gearbox, called a final drive. Differential steering subjects the machine's drive train to extreme torsional forces, so the final drives must be extremely strong and durable. This translates into big, heavy parts, which in turn translates into high cost. Taking a machine with wheels and putting it on full tracks instead would add at least 30 percent to the price of the machine. That's largely why previous attempts to put conventional tracks on combines had been commercial failures: Differentially steered tracks, with their attendant costs, priced the machines out of the market.

With the half-track system, by comparison, both tracks always moved in unison. The two tracks didn't need separate final drives, and no torsional forces resulted from their moving in opposite directions. Eliminating differential steering reduced the cost of the drive train by a considerable amount.

Neither Mark nor Ralph had seen a drawing of the "tippy-toe" half-track system, but they could imagine it. It struck them as being distinctly odd.

"We need to come up with a less expensive track system if the combine is going to be competitively priced," said Ralph. "But this tippy-toe deal isn't what we had in mind."

"No. It's sure not what *I* had in mind," Mark said, shaking his head with emphasis.

"So far we haven't been willing to make major compromises in Mark's design," Ralph continued. "I think that's why we've gotten as far as we have. We want to go with Cat, but we don't want to go down the wrong path."

"One of the guys who's coming tomorrow invented this tippy-toe system," Mark said. "It's his baby, so we've got to tread carefully. We don't want to step on his toes." He was silent for a moment, staring at the fire. He knew what it was like to have people dismiss your inventions without giving them an open-minded hearing. Worse, he knew what it was like to be treated with condescension. "We gotta watch our step," he repeated.

The next morning dawned with leaden skies and a light pelleting of snow, with heavier accumulations predicted. Downtown Des Moines looked gloomy despite the ersatz candy canes, snowmen, and fir boughs that hung from every light pole. At 9:00 a.m., Ralph and Mark walked into the appointed hotel and shook hands with Bob Petersen and the inventor of the "tippy-toe" system, who will here be called Dean Bruscher. Everyone went to a room on the fourth floor and took a seat at a round table. After a few pleasantries, Ralph launched into his pitch. "If you're going to get into the combine market," he told the two men from Cat, "you've got to come out way ahead of John Deere and Case-International. You can't just put tracks on existing technology and paint it yellow, or you'll fall flat. We see Mark's package of innovations as the best way for you to get around that."

Petersen and Bruscher agreed. They were both about fifty years old, and each had a full head of hair on its way toward silver-gray. They were dressed informally in jeans and shirts with open collars, and were similar in bearing—soft-spoken, friendly, distinctly engineerish. Bruscher asked Mark about his work on the self-leveling sieves. In the course of his reply, Mark pulled from his briefcase a copy of a John Deere patent that had surfaced during a patent search. The patent described a device intended to improve sieve performance on sloping ground. This was precisely the intent of Mark's self-leveling mechanism, although the approach shown in the Deere patent seemed more cumbersome. It required the addition of an extra auger and return gutter to the machine.

Bruscher examined the patent. "This is an add-on device to cure a deficiency upstream," he said. "That's not the way to do it. It's a Band-Aid solution. We're familiar with Band-Aids, too." He grinned knowingly at his colleague.

Ralph said, "I think John Deere looks at the Bi-Rotor and they think, 'How can we take parts of this and retrofit them on our machines?' They're not seeing the total picture of a fundamentally new and simpler combine."

"Well," Petersen said, "as we know, it's difficult for a big company to start with a clean sheet of paper."

"That's right," Bruscher said. "It's kind of a universal rule: A new machine can have no more than twenty percent new content. But then the whole thing becomes one big, giant Band-Aid. What you guys are doing is starting from scratch, and doing things right from the start. In a way, we're lucky we don't have a combine out there already to constrain us. We can start with a new center line. The established players in the market can't do that."

Mark unfolded some blueprints of the Kincaid prototype, drawn up before Cat entered the conversation. "The whole thing will have to be modified for tracks," he explained, "but this'll give you an idea of how it all fits together. The sieves are only part of the package. Here you can see my new elevation system, and the paddle-type unloader. We're in the process of patenting these systems. There's only one conventional auger on this whole machine, to distribute grain across the bin. There's an unbelievable amount of mechanism on most combines just in the unloading system. We've eliminated most of it."

The Cat engineers pored over Mark's drawings.

"Mark is always thinking not only about improving performance but also serviceability," Ralph chimed in. "He's asking himself, 'What has always bothered me about these machines, and how can we fix it?'

We like to say that the Bi-Rotor is designed by a farmer, with the farmer in mind."

"But who is designing the weld joints?" Bruscher asked. "That's important for how these things hold up over time. All machines have natural-frequency vibrations—those hums and rattles that crop up at certain rpms. Those are the things that kill you over time. They stress out the metal with fatigue."

"And is anybody engineering for single-event loads—things like driving over a ditch?" Petersen asked. "We can't tell you the best way to configure this combine, but we've got some expertise on designing for durability, to prevent fatigue problems down the road."

"I've observed those natural-frequency vibrations and seen their effects," Mark said. "I have a seat-of-the-pants idea of how to solve them. But as to how you calculate those things, I have no idea."

"You can't work with inventors who won't budge from their original ideas," Ralph said, nudging toward a new subject. "We've changed some things as a result of people making suggestions. I don't think we're hardheaded. But on some essentials, as we went along, we weren't willing to let others mess with them."

"Well," said Petersen, "once you start moving down the assembly line, it's all compromise." He laughed.

This brought the discussion around to the delicate issue of the tracks. Bruscher pulled out several large drawings of his half-track system, in plan and section.

"Steering is the problem when you introduce tracks," he began. "On full tracks with differential-speed steering, torsional forces on the final drives multiply. You have to have very heavy final drives, which adds cost. A track/wheel hybrid like this is an expedient."

Everybody looked at the drawings. They depicted something that looked a bit like the undercarriage of a World War II–era military half-track vehicle. The main difference was that the steering wheels in Bruscher's design were in the back instead of the front, and they moved up and down. Bruscher explained his system, emphasizing its benefits.

Mark and Ralph looked at the drawings, tight-lipped. Their apprehension did not go unnoticed.

"We recognize the visual-image problem of mixing tracks and wheels, and we don't like that," Petersen said. "But to meet your time schedule of coming out next harvest, and to meet the cost requirements that I think we'd all like to see for this machine, this is what we have to go on right now. It's a first iteration."

Bruscher: "It just comes back to: To steer a two-tracked vehicle is expensive. The half-track system involves compromises, but we're still trying to figure out how to steer a full-track machine inexpensively. This kind of system is an order of magnitude less expensive than full tracks. It would cost maybe a third as much."

Ralph: "How maneuverable and responsive is this steering system?"

Bruscher: "There's a bit of a time lag in response. But you get used to it. It's like with a normal tractor: You steer it with the front wheels, and if it doesn't respond right away, you use the differential brakes on the rear wheels."

Ralph: "Hmmmmmm. Sounds like an inexperienced operator might tear out some fence on the end rows."

The meeting went on like a tennis game. It was a polite game, though. Nobody tried to make points with slams or topspin. Rather, all shots were placed where they could be returned with a minimum of heavy breathing. A sense of wanting to please one another prevailed. Finally Bruscher said, "This has to be an iteration process, a feedback loop. You go back and think about it, and we'll do the same. This is a starting point."

That night, back by the wood stove, the gears were turning in Mark's head. He was quiet, staring through the glass doors of the stove into the flames. After a while he started sketching a track system on a scrap of paper.

"I'm already thinking," he said. "By the end of the week we'll fax them something to look at. I'm just thinking of how you could marriage this thing together and have a hell of a machine."

SOME WEEKS LATER Mark was taking his morning constitutional—"sitting on the pot," in his words. Suddenly he experienced a eureka moment. "I know! I'll double-transmission it," he thought. In his mind he saw a solution to the track quandary that by now was holding up progress on the new prototype. The solution, as he envisioned it, would satisfy all the requirements of low cost, durability, and simplicity. It was a full-track, differentially steered power train. But instead of using the huge, expensive final-drive mechanisms found, for example, on Cat Challengers and bulldozers, Mark's system would employ two of the smaller transmissions that were used singly on conventional, wheeled combines. Each of the two transmission gearboxes in his system would be powered by its own variable-speed hydraulic motor. The two gearboxes would be connected by a common axle, and one of the drives would have its differential disabled. This system would not be as heavy-duty as the drive train on a Cat D9 dozer, but it didn't need to be. Mark thought it would be more than adequate for a combine.

He sketched his idea on paper and faxed it to Peoria. The response was subdued but not dismissive. The world's largest manufacturer of track-driven machines had its doubts about this scheme dreamed up by a Kansas farmer. Armies of Caterpillar engineers—some of the best in the country—had designed scores of track systems over the decades. Many of them were similar to Mark's design in concept, although none was exactly like it. The Cat people weren't sure the Underwood system would hold up over years of hard service; they harbored doubts that it was worthy of the Cat imprimatur. But it would work, they could see that, and—its chief virtue for a prototype—it could be assembled from existing parts. They thought it was worth trying. They said, in effect, "Let's build it and see what happens."

On a bright, brisk February day, a semi pulled into the Gordon-Piatt building where Mark and Ralph were working. The truck had the yellow-and-black Caterpillar logo. A slightly plump, white-haired man jumped out.

"It's Santa Claus," Mark shouted.

Santa Claus—here to be known as Dick Dinkins, a jack-of-all-trades who worked for Cat's farm-equipment unit—had brought a sleigh full of toys to Kansas. Parked on Dinkins's flatbed was a Cat Challenger tractor, used but fully operational. The Challenger was to be dismantled and cannibalized for parts (chiefly its engine). Also on the flatbed were two enormous black loops of rubber. The loops were about two inches thick and three feet wide, and they formed circles roughly eight feet in diameter. Encased within the rubber, invisible to the eye, were reinforcing belts and cables of braided steel. The tracks, I would learn, weighed 1,800 pounds each. Near them on the flatbed lay two assemblies consisting of long rows of small steel wheels. The assemblies looked rather like a pair of giant Rollerblades minus the boot.

These ten-foot-long skates would be bolted to the sides of the prototype's chassis, and the tracks would fit over the rollers.

There was, however, no chassis to accept the tracks yet. Two months after the cousins arrived at Gordon-Piatt, they had yet to produce any physical sign of a new machine. Mark was still designing it. He was working with one of Gordon-Piatt's computer specialists to draw the combine with a CAD system. The same process had driven Mark to distraction at Kincaid, and he had ultimately abandoned the effort. Workers at Kincaid ended up fabricating combine parts from Mark's pencil drawings, sketched in some cases on the back of the fabled envelope. But envelopes wouldn't pass muster at Gordon-Piatt. The heart of the Gordon-Piatt plant was a computerized laser cutter and punch press—a steel-cutting robot. The robot could do some amazing things, but it couldn't read sketches on envelopes. It only understood instructions that were digitally encoded on magnetic floppy disks. The robot ate floppy disks for breakfast, and by lunchtime it coughed out stacks of perfectly cut sheet-metal parts.

Since Mark didn't speak the robot's language, he needed a translator. His translator at Gordon-Piatt was an excellent computer draftsman who knew the robot's language like a native speaker; the trouble was, he and Mark didn't communicate so well. It didn't help that the two men worked in separate buildings, or that the CAD specialist was available to help the cousins only a few hours a day. The result was that the project slowed to a crawl. Finished parts trickled back to the north building even more slowly than floppy disks trickled to the robot.

At this rate, Ralph realized, it would take them years to build a combine. He felt a sickening sense of déjà vu. Once again spring had arrived, the wheat was maturing, he was pushing to finish a combine by wheat harvest, but there was no combine in sight. The cousins' situation was exactly as it had been at Kincaid a year earlier. But the combine project was like a shark: It had to keep moving or it would die. Patents are only valid for seventeen years, and some of Mark's patents on the Bi-Rotor system were already more than five years old. Every year of delay before the machine made it to market was another year's profits lost. Plus, they needed test results before Cat would commit completely and offer them the grand payoff. If they missed the wheat again this year, they couldn't prove the machine in wheat for another twelve months—a dire setback for the cousins, and for Cat. A clock was ticking. The clock was attached to a time bomb that could eventually blow up the dream, not with a bang but a fizzle.

A modern combine—even the simplified one Mark had in mind—consists of some 30,000 components. Assembling that many parts in three months would require the addition of an average of twenty-eight components to the machine every hour for twelve hours a day, seven days a week. Not even Mark, with his exceptional shop skills and stamina, could do that. He would need a crew of assistants with various talents and areas of expertise, from welding to electronics to hydraulic systems. But to hire a crew, the cousins would need money, which, as of mid-January, they still didn't have. Bob Petersen knew that Mark and Ralph couldn't live on donated hardware alone. He promised he would send a cash allowance as soon as he could move a request through accounting (now that the calendar year had changed, he had a small discretionary account he could draw on). Ralph took Petersen at his word and proceeded on faith. He borrowed $20,000 from an investor to cover a few months of payroll expenses, then began recruiting.

The cousins' most pressing need was to speed up the CAD process. Until the combine was drawn on the computer, there could be no floppy disks to feed the robot; and until they fed the robot, there could be no combine parts. Ralph decided to hire a full-time CAD specialist—preferably someone who could climb inside Mark's head and pull out the combine, with forceps if necessary. Ralph could think of only one person who might be able to do that. His name was Glen Jackson, and he worked in the Texas shop where the cousins had built Whitey. True, Jackson was an engineer, and Mark was wary of engineers, but the two had gotten along extremely well in Texas. Ralph called Jackson one day and offered him a job. "I can only promise you three months of work," Ralph warned. "After that, I don't know what will happen."

Almost instantly, Jackson accepted. "I'll let you know when I can be there," he told Ralph.

Glen Jackson was elated. He was a lanky man in his mid-thirties with a narrow face, black hair and mustache, and gentle dark eyes. At home that evening, he told his wife, Patti, about Ralph's offer. She could tell Glen had made up his mind to take the job. It would mean he would have to live in Kansas during the workweek, leaving Patti, a schoolteacher, at home with the couple's two young children. But she took the whole thing philosophically. Glen, she knew, had been unhappy in his job. He had recently had a promising interview at Peterbilt Motor Company, the truck maker in nearby Denton, Texas. Maybe Peterbilt would offer Glen a position in a month or two; in the meantime, the combine project would be something new and different—a summer fling, of sorts, that Patti knew her husband would enjoy.

To Glen, it seemed a momentous decision. "I kind of felt like I was jumping off the edge of the world," he later recalled. "I was quitting a steady job near home to join a speculative venture four and a half hours away. To some people it might seem like a slightly crazy thing to do. But I had really enjoyed working on Whitey. Mark and I could read and interpret each other very well. Neither of us is a very good communicator by ourselves, but together, somehow, we clicked. We'd bounce ideas off each other and grow the thing. I was excited about what he and Ralph were trying to do, so I jumped at the chance to be a part of the project again."

The rest of the team fell into place, one by one.

Jeff Hawkins was a self-taught welder who had been working at Kincaid when Mark and Ralph were there. All his life Hawkins had been captivated by the romance of the wheat harvest, which was the most important event of the year in his hometown of Pretty Prairie, Kansas (population: 601). At harvesttime the twenty-five-year-old ex-football player with a walrus mustache would put in his nine hours at Kincaid, drive to a friend's farm after work and operate a combine until midnight, then get up at six the next morning and do it all over again. Mark's Bi-Rotor system made intuitive sense to Hawkins as he learned more about it. He took a particular interest in the project, spending as much time hanging around Mark as he could. When he found out that the cousins were leaving Kincaid to take their project elsewhere, Hawkins cornered Ralph. "You keep saying that a person only gets one or two chances to live out their dream," the young welder said. "Well, this is my chance. I want to *make* something of my life, not just punch a clock for the next fifty years. You've got to hire me. You've just *got* to."

Ralph found Hawkins's argument impossible to refute. Now that he had some money with which to pay him, Ralph hired Hawkins (with apologies to Delmer Kincaid). "For me this is the chance of a lifetime," Jeff said when he arrived in Winfield. "This is a chance to make history."

One day soon after he had hired Jeff, Ralph received a résumé in the mail, unsolicited. It came from Alan Van Nahmen, who lived in Columbus, Indiana. If Ralph and Mark had had a fan club, Van Nahmen would have been a charter member. He had first read about the cousins in a farm magazine a few years earlier and had clipped the article and saved it. "These guys are on to something," he later remembered thinking. Van Nahmen was in a position to know. He was a veteran of the farm-equipment industry with a special interest in combines. He had worked for John Deere for twelve years, and had lately been in charge of marketing track-equipped Claas combines in the United States. Tracks, he believed, represented the future of farm equipment. He was an amateur pilot, and often flew over Indiana farm fields in small planes. Looking down, he could see the tracks that the heavy tractors and combines made long after they had passed. The crops were stunted and discolored in their paths. He realized that this was happening to the best soil in America—the soil was systematically being ruined by heavy machines. To this problem there were two possible solutions, he thought: Either the machines would have to go back to being smaller or they would have to be mounted on wide, buoyant tracks so they didn't compact the soil so badly. He thought it unlikely that farm machines would shrink, so he became a zealous devotee of tracks and other technologies that would help farmers make better use of their precious ground.

Alan Van Nahmen ran into Mark and Ralph at a soybean convention during the winter of 1992 and learned, confidentially, that the cousins' new Bi-Rotor prototype might be mounted on tracks. Now *that* was interesting, Alan thought—this could be the machine of the future. A few months later he sent his résumé to Ralph's address in Fort Worth. One of the many job descriptions on the densely typed, three-page résumé was "factory school instructor for electrical, power train, and hydraulic functions" at the John Deere Harvester Works. Ralph liked the sound of this. He invited Alan to join the Bi-Rotor crew as its specialist in hydraulic and electrical systems.

Columbus, Indiana, was 740 miles from the Gordon-Piatt plant—too far even for weekend commuting. Even so, Alan took leave of his wife and two sons (just for a few months, they all thought), hopped into his silver-and-black pickup, and drove twelve hours straight to show up for work.

Glen, Jeff, and Alan were soon joined by a final team member, Sushil Dwyer, the Ph.D. candidate who had helped Mark and Ralph do their testing at Kansas State. In a previous life (or so it seemed) Dwyer had been dean of agricultural engineering at the University of Allahabad, on the fertile Ganges Plain in north-central India. By the age of forty-five, Dwyer had attained the pinnacle of success in his field, but found little satisfaction in this accomplishment. His duties were entirely administrative, yet he ached to do research, to invent. Ideas smoldered inside him, burning to get out. As a practicing Christian in an overwhelmingly Hindu academic and governmental bureaucracy, however, he felt he had little hope of getting a research grant. Even if, by some miracle, he was able to build one of his inventions, Dwyer believed that he would receive none of the credit or reward.

Dwyer's greatest interest in life was mechanical threshing, especially of wheat. Uttar Pradesh, the state in which Allahabad is a major city, encompasses the heart of India's wheat belt. Most of the crop is harvested by villagers with scythes, then threshed with stationary separators much like the ones that were used in the United States a century ago. Dwyer's mind tumbled with ideas about more modern threshing methods. He was a self-styled "harvesting man," from hull to germ. While carrying out his administrative duties, he dreamed of concentrating all his energy on the problems of threshing. The United States, he believed, was the one place where he could pursue this dream. There his ideas would be judged on their merits, regardless of his religion.

In 1984 Dwyer's wife of nineteen years, Shakundala, died of cancer. The couple had two sons, both nearly grown. Once the boys left home for college, Dwyer decided it was time for a change. He applied to doctoral programs at several American universities known for their agricultural engineering programs—Kansas State, Texas A&M, Purdue, Penn State, Iowa State. Kansas responded first, in the affirmative. Kansas was one of the world's leading centers of wheat research, so Dwyer packed his bags and moved from the chapati basket of India to the breadbasket of America. He left New Delhi on January 3, 1989, and arrived in New York twenty-four hours later, on January 3, 1989. "I lived through that day twice," he said later—"once in my old life and once in my new life."

As of the spring of 1994, Sushil had been back to India only once, in August 1989. He returned in order to marry a woman whom he had known for twenty years, Avis Dromila. Like Sushil, Avis had two college-aged sons. The newlyweds lived together for two weeks, then Sushil returned to Kansas. He has not seen his wife since, though he calls her often, spending an average of $500 a month on phone bills. He hopes that someday she will be able to join him in Kansas.

When Mark and Ralph showed up at Kansas State with their proposal to test the Bi-Rotor threshing concept, Sushil was just getting under way with his doctoral studies there. No one in the department of agricultural engineering thought the idea had enough merit to pursue—no one except Sushil. He approached his supervisor, who was the chairman of the department, and said with certainty, "This will work." The chairman looked at him skeptically and said in essence, "Okay, go prove it." Sushil's advocacy and assistance made it possible for Mark and Ralph to do just that. The three of them built and tested the lab prototype, proving not only that the Bi-Rotor idea worked but that it worked extremely well. Suddenly the skeptics at Kansas State began singing the Bi-Rotor's praises, claiming Mark's invention as a feather in the university's cap.

No one was more conscientious about returning favors than Ralph Lagergren. "Sushil," said Ralph at the conclusion of the testing, "if I can ever afford to, I'll give you a job." He made good on the promise in February 1993. Sushil had just finished his doctoral program. He collected his degree, climbed into his battered Ford Fairlane, and drove from Manhattan, Kansas, to the unmarked, curry-colored building at Strother Field. At long last, the slightly pudgy, white-haired "harvesting man," his high-pitched English heavily accented with the Hindi of Uttar Pradesh, had found a milieu in which he could see some of his ideas about threshing come to life.

By THE END of March the team was assembled. Mark and Glen put the finishing touches on the machine's overall layout—its gross shape and dimensions. Then, on his computer screen, Glen tore the combine apart, breaking it down into component parts. He worked in a makeshift office at one end of the Winnebago, which was parked in the middle of the shop floor. Glen had to draw each part on his computer screen, specifying its measurements to the nearest two-hundredths of an inch. When he finished entering a few parts in the computer, he would load the information onto a floppy disk and take it over to the Gordon-Piatt plant. There somebody fed his disks to the robot. The robot spat out flat pieces of sheet metal, cut to Glen's specifications. Human metalworkers deburred the pieces and bent those that needed bending into final shape. Then a member of the combine team—usually Ralph, Alan, or Jeff—picked up the parts and took them back to the north building.

The Gordon-Piatt fabricators who made the parts had no idea what the parts were for. Caterpillar had taken a sudden interest in maintaining a mantle of secrecy around the new prototype. Large corporations instinctively breed a culture of secrecy, and now the north building at Gordon-Piatt became a colonial outpost of that culture. Mark was glad to oblige in that regard; his instincts, too, counseled seclusion and privacy. He was fiercely protective of his brainchild, to the point where all outsiders had become suspect. Once when I pulled up to the north building in a new rental car, Mark ran toward the car brandishing a steel bar. When he saw it was only me, he acted as if the near-attack was a joke. I wasn't so sure, though, that his ferocity was entirely feigned.

The team laid the first lines of steel for the machine in April. On a typical day, Glen sat at his computer in the Winnebago, detailing parts. Jeff worked like a tinker at a big metalwork table, welding, hammering, assembling parts. Alan spent a good deal of time with his nose buried in suppliers' catalogs, finding and then ordering the correct hydraulic pumps, motors, switches, wiring harnesses, and hundreds of other components. Sushil's purview was the Bi-Rotor mechanism—the rotating cylinder and cage at the heart of the machine. Working at a drafting board, he drew the entire assembly, then broke it down into manufacturable parts. Mark moved here and there conferring with various team members, sometimes welding next to Jeff, at other times drawing parts on the concrete floor with chalk as he brainstormed with Glen. Ralph did whatever he could to help, tidying up with a broom or putting newly arrived boxes of nuts and bolts into the cubbyholes of two tall metal cabinets. On one of these cabinets someone had stuck a crudely lettered cardboard sign that said, "Ralph's Hardware."

By late May, the combine's basic frame was mostly complete. The new prototype now resembled the old, abandoned one. Both of them looked like heavy-duty jungle gyms propped up on blocks. The team had accomplished in a month what it had taken the cousins a year to do alone. They had gotten a slow start on the new machine—XBR2, they called it: for "Experimental Bi-Rotor number Two." With a willed naïveté that bordered on self-delusion, they convinced themselves that they could still finish in time for the wheat harvest. Much evidence suggested otherwise. On the floor outside the Winnebago, a few plugs of soil sprouted wheat stalks. The plugs were crop samples that Glen had brought back with him after weekend trips home to Texas. The wheat in the plugs was ready to cut. The line of ripening, working its way gradually north, would hit Kansas in a few weeks. Many of the fields around Winfield had already turned from green to gold. In a month, custom harvesters would arrive like a cloud of locusts, devouring wheat fields with their combines.

Eight days later the team had made little visible progress on the frame. Delays in getting parts back from the plant were holding things up. Sometimes a week went by between the time a CAD disk went into the plant and the completed parts came back to the north building.

On a cool, cloudy Wednesday morning in the first week of June, Jeff Hawkins stood at his worktable, bent over a large metal frame. His arc welder threw off ten-foot rooster tails of sparks as he fabricated the shoe assembly—the frame that would hold the sieve panels as they shook back and forth behind the rotor. A radio, permanently tuned to a country-western station, blared in the background. Jeff finished the weld and pushed his visor up. I stood nearby, admiring his work. "Do you really think you guys are going to cut wheat with this machine in a couple weeks?" I asked.

Jeff nodded, slowly so his visor wouldn't flop down. As usual, he had a plug of Copenhagen chewing tobacco tucked behind the left tusk of his walrus mustache. "We'll have the rotor in next week," he said. "In two weeks, the engine will be in and running. From here on out, this machine will come together fast. Will we cut wheat? I'd say most definitely yes."

I asked the same question of Ralph. "Yeah, we'll make it," he said. "We've got to. Tucker built his car prototype in sixty days. He didn't have a lot of people either. With teamwork you can do a lot."

That night I ate supper with Ralph. We went to the Mule Barn, a dark, smoky tavern and supper club in Ark City, where we sat and talked over beer and steaks. George Strait, Reba McEntire, Tanya Tucker, and Billy Ray Cyrus sang about cryin', lovin', and leavin' on the jukebox. Despite his can-do attitude, Ralph looked nervous, maybe even a little grim.

He had had a difficult spring. From February to April, while recruiting the team, he had also been negotiating two important contracts. His opponent across the negotiating table was Bernard Taylor. One of the contracts they hashed out was intended to formalize the relationship between Agri-Technology (that is, the cousins and their original investors) and Gordon-Piatt. Taylor, as a manufacturing representative, acted as Gordon-Piatt's agent in this matter. The other contract spelled out an agreement between Agri-Technology and a corporate entity called Paramount. Paramount was a newly created business partnership between Taylor and the two cousins. In the scheme they had all worked out, Paramount would assemble Bi-Rotor combines from parts made by Gordon-Piatt and other fabricators. In this negotiation Ralph represented Agri-Technology and Taylor bargained on his own behalf, as a member of the Paramount partnership.

Bernard Taylor had previously treated Ralph with fatherly solicitude, but at the negotiating table he became as tough as an old rooster. Ralph felt disoriented by this change. He had grown accustomed to regarding Taylor as his avuncular ally. Also, Ralph instinctively avoided conflict. His talents as a salesman predisposed him to find the quickest, surest route to agreement and bonhomie. He read people quickly and made a beeline to their good graces.

But the negotiations required something different of him. Bernard was aggressive, angling stubbornly for the best possible deal. He was "only taking care of business," Ralph recalled. "It was nothing personal, but it was still hard for me." With Scott Larson backing him up from Dallas, Ralph held his ground. He was determined to protect his original investors. ("It would be easy for me just to let them get screwed, but I won't do that," he said.) Ralph was even more determined to make sure that, when all the dust settled, he and Mark still had something left for themselves. "If we agreed to all the concessions that people have asked of us over the years," Ralph told me, "we'd have lost the whole thing by now.

"I mean, here we were on the final runway," he said. "We've been at this for ten years. Now we have Cat behind us, and I was getting a good team together. Our goal was finally within reach. But I was at a point in the negotiations where I could have fucked it all up in an afternoon. You can lose it all in one day as an entrepreneur."

Taylor, an old hand at hard bargaining, took the negotiations easily in stride. "We never argued about a damn thing," he said later. "Ralph was liberal as hell. The contracts we came up with are fair for everybody." Now that it was over, he and Ralph went back to being friends and allies. But the months of wrangling took a toll on Ralph.

All that remained after Ralph and Bernard finished negotiating was for Gordon-Piatt to review and approve the details of its part in the deal. Ralph had expected this step to go smoothly. The worst, he thought, was over.

"But then came this afternoon," he told me. "I got a call from Jim Salomon. He had just had a meeting with the Gordon-Piatt board—a six-hour phone meeting, and the main thing they talked about was us. The board members don't like our contract. What it boils down to is, they want exclusive manufacturing rights."

He took a bite of filet mignon, a swig of beer, and shook his head. "I'm happy to give them plenty of manufacturing. They deserve it. But we can't make it *exclusive.* That would tie our hands too much. What if, a year down the road, they started doing a sloppy job? Or their prices got too high? Or Cat decided they wanted to do the manufacturing in-house? Gordon-Piatt is a good company, but they have to prove to us that they're the right people for the job. If they had exclusive manufacturing rights built into the contract, they'd have no incentive to perform for us."

"So what's going to happen?"

"There's a meeting in a couple of weeks. Me, the Taylors, and Jim Salomon. If we can't settle this, Mark and me and the guys could wind up on the street. Gordon-Piatt could just kick us out, like, 'Have your stuff out of here by morning.' To move out of the shop, with the deadlines we've got? That would be hell."

Gordon-Piatt was owned by six investors. Two of them worked for the company: One was vice president and chief operating officer, and the other was Jim Salomon, who had the titles of president and CEO. Years before, under different management, Gordon-Piatt teetered on the brink of bankruptcy. Salomon, who was then plant manager, lined up some wealthy investors and organized a buyout. Then, as president, he turned the company around. In 1992 it won a Subcontractor of the Year award from the U.S. Small Business Administration. Salomon provided the brains and chutzpah behind the company's rebirth, while his four investors, who lived in Florida and New Jersey, provided the money. The investors rarely set foot in Kansas, but they had a large say in how the business was run. It was they, not Salomon, who insisted on exclusive manufacturing rights in the combine deal.

"They've been successful," Ralph said of the absentee owners. "Their bellies are full. They're not hungry."

Jim Salomon was stuck in the middle. Though he couldn't say so, he probably would have been satisfied with the contract that Ralph and Bernard had so strenuously hashed out. He wanted the deal to work, hoping that he'd have a nice little combine factory humming alongside his burner factory someday. But he was obliged to represent the position of his board members, who didn't seem to care much if the XBR2 and its crew stayed or left.

"Next month we'll still be here, and we'll finish it here," Ralph said with a determined look. "That's my job."

We finished our steaks. Ralph's mood had improved. He looked more relaxed. The old smile came back. "Oh, by the way," he said. "I'm engaged. It happened a month ago, but we kept it quiet until now. I didn't even tell Mark until a few days ago. We want to get married this fall."

"You're *what?*" I shouted, slamming down my mug. The tattooed crowd standing around two nearby pool tables looked over at our booth, perhaps anticipating a fight. But, seeing Ralph's broad grin, they went back to playing eight ball. "You're engaged?" I said, more softly now. Ralph savored the moment, watching my reaction. "Okay, so who's the girl?" I finally asked.

"Her name's Dawn. I met her this winter, out dancing. She lives up in Lyons. After her dad skipped out on her family, she and her brother and sisters were supported mostly by their granddad. So this last year, out of appreciation, Dawn saved up and bought her granddad a new truck. I couldn't believe that. She bought her granddad a pickup out of what she earned waiting tables at Pizza Hut. You don't meet too many people like that. Plus, she's the best dancing partner I ever had."

"Well, here's to good dancing partners—to you and Dawn," I said, raising my glass.

"And to the Bi-Rotor," Ralph added. We clinked mugs.

• • •

Two weeks later, on a Thursday morning in mid-June, a dark-green Cadillac pulled up to the curb in front of Gordon-Piatt's main office. Bernard Taylor got out of the car slowly, aided by his son, James. Ralph was waiting for them. It was the day of the big meeting with Jim Salomon. Ralph was about to find out if the Bi-Rotor team could stay at Gordon-Piatt or if they were going to be kicked out. He had not informed Mark or any of the other guys of this possibility. Ralph didn't want them to worry about it. He was buoyant—even a little giddy—primed to make his pitch. The three men disappeared into Salomon's office.

Ralph had briefed me on his strategy that morning. "I'll tell Salomon that we can, if we need to, just pay Gordon-Piatt for their work on the prototype out of our own pockets. We'll pay rent for the use of the shop, we'll pay for parts. Then, when it's time to manufacture combines, they can bid for fabrication work just like anyone else. Or they can accept the deal and be guaranteed a major share of the work. I'll tell him we've walked away from other people when they started making unreasonable demands. I'd hate to see that happen here, but it could."

Salomon responded to this speech, I soon learned, with a repetition of his board's position: We respect what you're doing, but we're playing an important role here and feel we're entitled to all fabrication work that comes out of it.

"Damn it, Jim," Bernard erupted. "This is a good deal, but your board can't see it."

In the horse-trading that followed, Ralph stuck to the basic structure of the original agreement but added a few sweeteners—escape clauses and enhancements that would reward Gordon-Piatt for its investment no matter what happened.

Finally Salomon said, "I think this is a fair deal. We're tremendously excited by this thing, and really do see the potential. I'll talk to my people again and see what they say."

Ralph, feeling the momentum moving his way, raised another issue. "We've got to move quickly on this machine to fulfill our obligations to Cat," he said. "But we've been slowed down because of delays in getting parts back from the plant."

Salomon picked up his phone and got the plant manager on the line. "Do we have these parts on high priority? No? Well, put them there. I want forty-eight-hour turnaround on these parts. Treat them as though there's a ten-thousand-dollar-a-day penalty for being late." He hung up and looked at the three other men in his office. "That ought to do it."

Before Ralph and the two Taylors left his office, Salomon voiced his own enthusiasm for the new machine. "I want to drive one of these combines out the door," he said.

Ralph thought it was time to kindle a little of the boss's enthusiasm for the project among the Gordon-Piatt workers, who still had no idea what was going on in the north building. He met with the union stewards in the plant and told them the Bi-Rotor story. His reasons for doing this were mainly selfish. He hoped that by removing the aura of anonymity and secrecy from the project, he could rally the workers to the team's cause. He hoped they would speed up the delivery of parts, not just because their boss had told them to but because they supported the cousins and wanted to participate in this expression of the American dream. Ralph gave some newspaper clippings about the combine to the union stewards, who, on their own initiative, made copies and distributed them in bathrooms around the plant.

The following week parts began coming back to the north building less than twenty-four hours after Glen delivered his CAD disks to the plant. Before, when members of the combine team went to Shipping and Receiving to pick up finished parts, they were acknowledged with expressionless nods; now they were greeted with enthusiastic smiles and inquiries.

Ralph knew that Bob Petersen and even Mark would probably disagree with his decision to spill their secret to the rank-and-file workers in the plant. He had definitely compromised the security of the combine project. Who knew, now, how far and wide the news of what they were doing might travel? But Ralph made a conscious habit of trusting his gut, and his gut had told him that the benefits of enlisting the workers' support would outweigh any problems that might result from a breach of security. Any determined industrial spy, he reasoned, would have found them long ago anyway. When he observed the new collegial spirit with which he and the others were greeted at the plant—and, more important, the new speed with which parts were coming back—he felt he had made the right decision.

No amount of speed at the plant, however, could alter one fact: They were not going to make the wheat harvest. The idea that they might have finished in time to cut wheat had been a necessary fiction that had kept them working hard all through the spring. Now, in late June, with combines plying the fields all around them and the prototype nowhere near completion, delusion no longer served a positive purpose. They had a new, even more critical deadline to worry about. Bob Petersen had just told them about a big corporate powwow that Caterpillar planned to hold in August. Top management and key dealers would get together near Peoria to evaluate new machinery models. Petersen and others wanted the combine to be there on display. Cat's management wanted to get some early feedback on the combine from the dealers who would be selling it. The impression the machine made in August would weigh heavily in Caterpillar's decision to carry on with the collaboration—or not to.

The team had scarcely two months to go until the Peoria deadline, but the machine resembled the skeleton of a beached whale. Could these bones live, and in a mere eight weeks? It began to look doubtful. Yet if the machine didn't make it to the Peoria meeting in presentable shape, Cat might give up on the Bi-Rotor boys—write them off as a bunch of country incompetents.

"You know how important August nineteenth is, don't you?" a Cat executive asked Ralph during a midsummer phone conversation.

"Yeah," Ralph answered. "We only have our whole life riding on it."

The team worked twelve- and fourteen-hour days, seven days a week, but the pace of progress still wasn't fast enough. Ralph concluded that the answer, once again, was to recruit more people. His hand was freed somewhat, since he had finally received a cash allowance from Cat. A check had arrived for more than $200,000, which sounded like a lot until you considered that nearly half of it had already been spent on components and hardware that neither Cat nor Gordon-Piatt could supply. Several expensive components had yet to be ordered and paid for. Also, Ralph was already writing checks for more than $15,000 a month to cover taxes, workmen's compensation, and payroll expenses. Adding to the payroll meant that the allowance from Cat would dry up in just a few more months. Still, Ralph felt he had no choice but to hire more workers.

His first summer recruit was Joe Lutgen, a newly minted graduate of Kansas State University. An engineering-technology major in college, Joe was bright, a hard worker, and a CAD whiz kid. His job was to assist Glen in drawing and detailing combine parts on the computer. Joe was the young brother of two of Ralph's investors, so he was, in a sense, family.

Ralph's next hire was a member of the extended Bi-Rotor family, too. Ralph was wondering how to find another shop helper for Mark one day when Glen Jackson happened to mention that his father, Ken, was temporarily out of work. Ken Jackson was a civil engineer who had supervised construction jobs all over the world. He had eventually settled down in Texas, as a maintenance manager in a factory near the home where he and his wife (Glen's mother) lived. Ken didn't relish this work, since the day-to-day problems he dealt with usually had more to do with personnel than with engineering. Even so, he did his job well and faithfully for many years. Then, abruptly, the factory closed and Ken was given early retirement. When Ralph suggested that Glen call his father to ask if he would help out on the combine for a while, Ken accepted with enthusiasm, though the pay would be modest.

Ken's greatest passion was making things in a shop with his hands. In fact, he was one of the handiest men alive. Mark could say, "I guess we need a little reinforcing here," or, "Somehow these rods have to be mounted on the cross-distributor chain," and Ken would do it with quiet competence. A week after he started working with the Bi-Rotor team, Ken received an attractive job offer back in Texas. He drove home that weekend, expecting to accept the offer, though he felt strangely sad about the prospect. Taking the job was what was expected of him, he thought, but it wasn't what *he* wanted to do. At the last minute, he decided to turn the job down. He drove back up to Kansas the next morning, put on his work apron, and happily returned to his lowly role in the shop. Ken bunked in a pop-up camper that he had towed from home and set up in a remote corner of the north building. In the camper, on a stand by Ken's bed, lay a single book. Its title was *Do What You Love, the Money Will Follow.*

A ninth worker joined the crew as a summer helper. He was Aaron McKee, Alan Van Nahmen's nephew. Aaron was between years as a mechanical-engineering major at Kansas State. He was glad for this reprieve from a summer of working on the farm in western Kansas, where he had grown up. "Things are kind of depressing on the farm," he said. "There's no living in farming, no matter how hard you work.

But on this project you can see what you've done every day. These guys are *doing* something. It's a lot more upbeat than farming."

The atmosphere in the north building was upbeat for the most part. Glen had a fine talent for leavening the long hours with good humor and the occasional pun. The later the crew worked at night, the worse his puns became. Ralph also worked hard to keep the mood light. He was always up—grinning, joking around with the guys, slapping them on the back. But the real jester of the group was Alan. He was usually serious on the job, then turned into a cut-up after hours. He had, for example, rigged a motorized squirter behind the grille of his pickup and amused his passengers by drenching unsuspecting pedestrians. He knew one particularly good card trick, which he sometimes performed in bars after he'd had a few beers. And on festive occasions he drank from a special beer mug that he kept in his pickup. The mug had a miniature marquee around its base that lit up with the revolving message "It's Party Time!" whenever Alan lifted it off the table.

June passed, then July. The days grew hotter and more humid as the August deadline approached. Strangely, the amount of work that remained to make the machine presentable seemed not to diminish but to grow. Projections about how long various jobs and steps would take almost always turned out to be wildly optimistic. Everything took at least twice as long as it was supposed to take. The long hours and seemingly slow progress began to take their toll. Habitual smiles disappeared, jokes and laughter subsided, friendly conversation ebbed, tempers frayed. All that could be heard for hours on end was the whining and pounding of tools, accompanied by the plaintive crooning of Garth Brooks and Wynonna Judd.

No one showed the increasing strain more than Mark. Over the summer months he became steadily more subdued and withdrawn. In July, when the push to complete the machine reached its peak, he went home to harvest 300 acres of wheat that he had planted the previous fall. The timing of his absence from the north building could hardly have been worse, but Mark felt under enormous pressure at home, too. One source of that pressure was his deteriorating marriage. More than three years of his living away from home had turned husband and wife into sullen strangers. The situation only aggravated Mark's fears about family finances. They owed money on the farm, which put pressure on Mark to make good on his wheat crop. But the nonstop rains that flooded fields and towns all over Iowa, Illinois, and Missouri during the summer of 1993 afflicted Kansas as well. Some of Mark's wheat had been drowned in all the moisture; a good deal of the rest was unharvestable because the fields were too muddy. After two weeks of trying, he managed to recover only about half of a normal crop. He lived in fear that the loan officers at the bank would call in his note and force him to default on the old home place.

Mark's fears were all the worse because he had sold most of his farm equipment the previous winter. He had hoped to pay off his farm debt with proceeds from the sale, but the auction had been disappointing. Now he still owed the major portion of the debt, but with the sale of his machinery he could no longer fall back on farm income to help him repay it. Ralph had repeatedly told him, "I burned my ships when I quit my job. Now you've burned your ships, too." With his farm auction, Mark had burned them. He had bet the farm on the combine deal.

That was what he worried about most. He felt more paranoid now than ever that something was going to screw up the whole thing. He felt almost certain that John Deere was going to steal his ideas, or that Cat would pull out suddenly, leaving him twisting in the wind. "I'm pretty exposed and vulnerable here," he told me one day soon after returning from Burr Oak after the disappointing harvest. "I could lose everything."

In his depressed mood, Mark had a dark foreboding about the combine's approaching debut in Peoria. The Cat meeting was to be such a buttoned-up affair that neither Mark nor Ralph was invited to accompany their machine. They would have to send it off without either of them to guard and protect it. Mark confided his feelings about this to Sushil one evening. "Mark is feeling very sad," Sushil told me later. "He's sorry to be sending the machine away. It's his baby, and he feels they are taking it away from him. He doesn't know what will happen to it there."

These stresses, worries, and sadnesses sent Mark into a deep funk. Normally quick to smile or crack a joke, he withdrew into a somber shell. Working on the combine no longer seemed to give him pleasure. He smoked more and more, and his habits of working, eating, and sleeping became erratic. The closer the combine came to completion, the lower his spirits sank.

The rest of the team noticed, of course, and with deepening concern. Mark was their guru, the creative wellspring of their shared endeavor. Increasingly, though, the guru was absent, in spirit if not in body. Mark had, fortunately, articulated enough of his vision to other members of the team so that they could carry it forward in spite of his doldrums. Glen stepped forward to fill the leadership vacuum that had developed on the shop floor. He became the de facto project manager, directing the rest of the team as the machine progressed. The others accepted Glen's leadership willingly, even gratefully. He exercised it quietly but effectively, with a light touch and a firm command. The experience called forth skills he didn't know he had. "I didn't figure on coming in and being the key person," he told me one day in mid-July. "It's kind of new to me, being in charge. But I like it."

As Glen's role expanded, Joe Lutgen rose to the status of lead computer jockey. Joe even looked rather jockeylike. He was short in stature, with a round face and sandy-blond hair receding a bit above the temples. Jeff Hawkins sometimes called him Opie, Ron Howard's character in the old Andy Griffith TV show, whom Joe resembled. Joe had interviewed with several companies after finishing college, including Trane, a large manufacturer of heating and cooling equipment. "When I came down here to talk to Ralph," Joe told me, "I saw the guys out working in the shop, in a kind of makeshift setup. Then he took me to meet Glen, and I thought, 'I'd be working in a Winnebago?' It all looked so *uncorporate.* I thought, 'What kind of deal is this? I don't think this is for me.' But as I drove back home, I realized it was an opportunity to do everything I'd been wanting to do—to help design a new product using my CAD skills. When I got here, I was amazed at how much they let me do right away. In most companies they'd sit me off in a corner and have me do little minor stuff. But here, before I knew it, Glen told me to draw and spec the whole grain bin."

Far from being "little minor stuff," the grain bin was major, big stuff. It was, in fact, the largest single component on the combine. The bin was designed to hold as much as 400 bushels of grain. As a measure of comparison, a Volkswagen beetle weighs approximately 1,700 pounds. Let's call this unit of weight one VW. The payload of the XBR2's completed grain bin would be thirteen VWs—nearly twice as many VWs as any other combine could carry in its grain bin. In terms of space, 400 bushels of corn occupy 520 cubic feet. That's about the same volume as a room eight feet square with an eight-foot-high ceiling—a space, in other words, the size of a Manhattan kitchen or a Tokyo hotel room.

Joe Lutgen had never designed anything that big before. The challenge was made more complicated by the bin's irregular shape. It had to fit precisely between the combine's cab and engine compartment, while straddling the threshing chamber, with a low sump on one side where the unloader would fit. Joe broke the polyhedron down into about two dozen steel panels that would be held together with welded seams and some 300 nuts and bolts. When it was finally assembled on the floor of the north building, the bin resembled an upside-down house sitting on its roof.

A large grain capacity was one of the XBR2's most significant features. Because Mark's design dispensed with about five VWs worth of moving parts, that weight and space could be devoted to grain instead of iron. Farmers would welcome the larger bin capacity, since it would allow them to unload the combine less frequently. Especially in corn with its high yield, unloading the combine into waiting trucks and wagons is one of the most laborious and time-consuming operations of the corn harvest. In an eighty-acre field of good corn, you can't combine a full round (that is, go from one end of the field to the other and back) without having to unload somewhere in the middle—with a normal combine, that is. But with Joe's apartment-size bin, you could cut a full round without stopping. Such a big load would cause a normal combine, with wheels and tires, to make deep, compacted ruts in the soil. The XBR2's wide tracks, though, would greatly reduce the compacting effect of its loaded weight. A loaded XBR2 would be like twenty-three VWs of iron and grain tiptoeing across the field on Cat tracks.

One day three weeks before the Peoria deadline, the time came to hoist the grain bin into place on the machine. Joe was nervous. He had measured and remeasured every dimension many times, then double-checked his computer drawings with Glen before sending his CAD disk to the laser-cutting robot. When the parts came back from the plant, they all fit together as Joe hoped they would, which was an accomplishment in itself; still, he worried that the whole thing might be off somehow. He had never felt the weight of such responsibility. Weeks of time and thousands of dollars would be lost if this huge, ungainly box of sheet metal didn't fit in its appointed space.

The whole team helped swaddle Joe's baby in chains, then they lifted it with a big ceiling hoist. The bin rose slowly, then moved sideways into position over the combine frame. As it did, Joe's heart stopped. He thought he'd made a horrible mistake. Suddenly the bin looked far too big. Joe felt like throwing up, but he just stood there, transfixed, as his monstrous blunder descended toward the diminutive combine frame. Suddenly the bin was on; the *Eagle* had landed. To Joe's amazement, it fit just as it was supposed to. The bin made the machine look as top-heavy as an ant carrying a potato chip. That wasn't Joe's fault, though. He had done his job. He sat down to give his adrenaline-pumped body a few minutes to absorb the relief he felt in his mind.

Jeff Hawkins walked over to Joe, slapped him on the shoulder, and said, "Well, Opie, it looks like you did it."

With two weeks to go, the crew hoisted the cab and engine into place at the front and back of the machine. They slid the Bi-Rotor assembly into a gaping hole under the cab. Now the recently hollow frame was filled with guts. It had a head (the cab) and an abdomen (the engine) on either end of its husky thorax (the bin). Once the tracks were mounted, the machine's proportions improved. For the first time, it began to resemble a combine, though one with an unusual pedigree. Mark brightened at the sight of it. "It's gettin' to look like quite a machine, ain't it?" he said, smiling broadly for the first time in weeks. The rest of the crew got out their cameras to commemorate a major benchmark of progress.

A week later, with just one week to go, the Bi-Rotor team still hadn't started the combine's engine. Would it run? No one knew. Would the tracks work as planned? A thousand things could go wrong, and it was almost certain that some of them would. Prototypes never worked properly in the beginning. Even the most experienced prototype builders, working in the best-equipped machine shops money could buy, assembling prototypes whose every detail had been mapped out in advance by platoons of engineers—even then, problems arose. Hoses leaked, gaskets blew, pumps mysteriously failed. Here, in the comparatively informal setting of the north building—with a crew that lacked experience in prototype construction, with much of the planning and engineering done on the fly, and with a supply of tools that would hardly have sufficed to equip an auto-repair garage—in this setting, problems seemed not only likely but inevitable. The crew would need time to work out the bugs, yet time was the ingredient that was in shortest supply.

"It's going to be real tight," Ralph said. It was the morning of Monday, August 9, ten days before the meeting in Peoria. "But we don't have a lot of choice. We've *got* to be ready."

A dispute had arisen over what "ready" meant, under the circumstances. Hauling the combine to Peoria would take two days, and a third day would be needed to set it up. Before the machine could leave the north building, though, it had to be painted. That would take a day, with another day allotted to let the paint cure. All this meant that the last day the crew could work on the combine before it was painted was the following Friday—Friday the 13th.

The machine would not have to cut grain at the meeting in Peoria. All it had to do was sit at the site of the secret meeting and impress a select audience of Caterpillar dealers and executives while Bob Petersen gave a talk explaining the XBR2's innovative features. It would help if the machine's engine ran and its tracks operated so it could move by itself. The team members all agreed that their main goal, before the XBR2 left, was to get its engine and tracks running properly. That much was beyond debate. But there was also the matter of exterior metal. Plans called for eventually providing the machine with a sheet-metal "skin"—panels that would give the segmented body a unified shape, a more pleasing appearance. A majority of those involved in the combine project believed that installing the skins could wait until after the machine returned from Peoria. Even the project's supporters at Cat had told the team, "Look, guys, don't worry about external metal now. You can do that later, after the meeting."

Ralph felt otherwise. He rejected the idea that the skins were trivial ornament. "I'm sorry, but that's wrong," he said. "The machine is going up there to make a first impression. First impressions are important. It would be like presenting a new Cadillac to the dealers without a body. Would the dealers say, 'Oh, great car, we can't wait to have it on our lots'? No. They'd say, 'Why are they showing us this thing?' "

To bring the rest of the team around to his point of view, Ralph offered a bribe. "Okay, guys," he said. "If we have it skinned and painted by Saturday night, we'll all go out dancing, and I'll buy the beer."

Nobody argued. Ralph mounted a crash campaign to get the skins drawn on the computer, then rushed the CAD disk to the Gordon-Piatt robot. A crew of metalworkers at the plant worked overtime to finish the skins and get them back to the north building. By Tuesday,

August 10, the skin panels sat in a pile on the shop floor, ready for installation.

A bigger impediment to free beer than the skins was the combine's hydraulic system. Nearly everything on the XBR2 would run hydraulically—the tracks, the rotors, the self-leveling mechanism, the unloading conveyor, the dual-path elevator, and other vital components. The extensive use of hydraulics was part of Mark's strategy for eliminating moving parts: Instead of transmitting the engine's power to far-away parts of the machine by means of many belts, pulleys, and drive chains, Mark planned to outfit the XBR2 with several compact hydraulic motors that would drive each component directly. Mark's machine was to have eight hydraulic motors, compared with the one or two typical of conventional combines. All of the XBR2's most vital functions would run off the energy in a pressurized liquid.

A hydraulic motor works much like a hydroelectric turbine. Deep in the bowels of a hydroelectric dam, a turbine rotates when water, under pressure from the reservoir above it, flows through a penstock to the turbine. The pressurized water presses against blades on the turbine, causing the turbine to rotate, thereby turning a generator. A hydraulic motor on, say, a combine is an enclosed turbine. Instead of having a penstock, the combine has hoses or pipes, which direct a fluid to the motor from a hydraulic pump. The combine's engine runs the pump, which forces hydraulic oil through the supply hose at pressures of 2,500 pounds per square inch or more. The pressurized fluid runs into the motor casing and through an orifice, which directs the flow against the teeth of a turbinelike gear. The gear turns, causing a shaft to turn, and the shaft performs useful work. The fluid, much of its pressure energy spent, returns via a low-pressure hose to a reservoir near the pump. The same fluid recirculates over and over. You can vary the speed of the hydraulic motor by using a valve, like a spigot, to increase or decrease the flow of fluid through the supply hose. In a machine like the XBR2, small electric motors open and close hydraulic valves, and these electric motors are controlled by switches in the cab.

Another important function the hydraulic system serves in a combine (or tractor, or bulldozer, et cetera) is that of operating hydraulic cylinders. A cylinder moves something, such as a combine header, a plow, or a bulldozer bucket, up and down or back and forth. Hydraulic cylinders harness a physical principle discovered by Blaise Pascal (and known as Pascal's principle) to multiply the pressure of the incoming fluid, generating fantastic forces. With a pump operated by an eight-horsepower lawn mower engine, a few strategically placed hydraulic cylinders can lift a house. The XBR2's diesel engine, by comparison, was rated at around 300 horsepower. It needed all those "ponies," as Mark called the mechanical herd, to run the half-dozen hydraulic cylinders and eight hydraulic motors on the combine. Switches and levers in the cab acted like cutting gates in a corral, directing portions of the thundering herd (translated into pressurized oil) to this task and that.

The XBR2, its superstructure aside, was essentially a complex network of pumps, hoses, valves, motors, and cylinders—in other words, an enormous plumbing job. To get it done in time for the meeting, Caterpillar sent an employee whom I will call Ted Schwitters, a hydraulics expert who had worked on many prototypes during his years with the company. He summed up his role with modesty, saying, "I'm here to help with the plumbing."

Plumbing headquarters was a piece of plywood supported by two sawhorses standing a few yards from the XBR2. The makeshift table sagged under the weight of coiled hoses and metal fittings of every size and shape—threaded hose ends, connectors, elbows, bushings, nipples, and plugs. Ted selected parts from this jumble to make dozens of hydraulic lines, which ramified throughout the combine like rubber veins and arteries. Alan, meanwhile, strung wires, like nerves, from the cab (the combine's brain compartment) to remote sensors, lights, hydraulic valves, and engine parts. The two men worked quietly, steadily, barely stopping to eat lunch, dripping with sweat.

On Tuesday afternoon, with three days to go, the temperature inside the north building exceeded one hundred degrees. Several fans stood around the machine, languidly pushing around the hot, humid air. At 3:00 p.m., Mike Elswick, production manager at Gordon-Piatt, walked into the building with a camera. Elswick had sweated blood to keep the combine's parts moving through the plant on time, and he had become one of the project's principal backers at Gordon-Piatt. "I thought I'd come and hear it running," he said to no one in particular. To me, he confided, "It'll be a big boost to morale when they get that damn engine running. But it looks like it'll be a while yet." Louder, so the others would hear, he added, "Probably a few days yet, with *these* jokers working on it."

Elswick's attention alighted on four steel tripods that stood near the machine. "Are those jack stands?" he asked. "What are you going to do, put it up on those when you start it?"

"Yeah," said Jeff Hawkins. He had just completed a weld on one side of the machine. "It would be kind of a drag if this thing suddenly took off backward and went through the wall when we start it."

Elswick nodded in agreement. "How are you going to lift it onto them? Wait—don't tell me, I know. Ralph's going to *talk* it up. He's just going to talk that damn thing up in the air." He took a few pictures and left.

Ted Schwitters finished making the last few hydraulic lines at seven o'clock on Tuesday evening. He took the hoses to a car wash to flush them out. Hydraulic systems are like Achilles—they perform superhuman feats, but they're temperamental and vulnerable to tiny flaws. A bit of grit in a hose, a wrong-size orifice in a motor, a gasket askew in a pump, and the whole system freaks out. Ted returned by eight o'clock with his flushed hoses, their ends carefully taped shut to keep out specks of dirt. The hoses were all installed by nine. The two big hydraulic pumps below the engine now sprouted hoses like octopus arms, going every which way.

Alan, who was now helping Ted, started looking puckish. Alan was a reliable barometer of excitement. When he began looking mischievous, something was up. "Now she'll start," he said with a devilish grin. "We'll put some engine oil in her, and she'll start."

Ralph, infected by Alan's eagerness, climbed up on the engine platform and poured antifreeze into the radiator. Alan and Ted hoisted a fifty-five-gallon drum of hydraulic fluid ten feet into the air. The drum lay sideways in a sling, a spigot on its lid poised to anoint the combine with oil. Alan and Ted fitted a hose onto the spigot and started filling the hydraulic reservoir with fluid. They emptied one drum and tapped a second one, transfusing golden-brown lifeblood into the machine.

Glen led a detail that was preparing the machine for skins. He, Jeff, Joe, and Ken attached angle-iron struts to the side of the grain bin. The struts would hold the skin panels in the proper position. Inside the grain bin Mark, Sushil, and Aaron worked on the dual-path elevator, unaware of the sense of anticipation that was building on the ground. By ten o'clock, the anticipation was blunted by fatigue. The hydraulic reservoir took a long time to fill. What's more, the hydraulic pumps had to be primed and the transmission gearboxes had to be filled with gear oil before anyone could start the engine. By ten-thirty, everybody was too tired to savor the moment of ignition. They all decided to wait and fire her up first thing in the morning.

Wednesday, August 11, 8:30 a.m.: I was wedged in a narrow crevice between the left track and the machine, holding a funnel while Mark filled the left transmission gearbox with oil. Part of my duty was to watch a telltale hole in the gearbox casing, waiting for oil to appear. After I'd lain sideways for several minutes in a most uncomfortable position, ninety-weight oil started to gurgle out of the hole and onto my Levi's. "Whoa!" I shouted.

"She took three gallons," Mark said. "Okay, let's do the other side." I crawled to the other side, feeling like a greasy spelunker. "We'll fire her up here pretty soon," Mark said as the right gearbox filled. "It'll help boost morale around here—mine included. We could use a sign of accomplishment like that."

9:15 a.m.: All was ready. Alan said to Mark, "You want to start it?"

Mark climbed into the cab. "Kill that radio," he yelled. Travis Tritt fell silent in mid-yodel. "Contact," Mark shouted. He turned the key. Ted gave a little jerk on a temporary throttle lever that was mounted just below the engine. The big, six-cylinder diesel sprang to life. It purred with a low, contented rumble. Ted sprang into motion. He ran around to check several pressure gauges that he had plumbed into various places in the hydraulic system. Two of the gauges registered more than a thousand pounds per square inch. That was good. But the needle on a third gauge quivered at zero. When Ted saw this he ran back to the throttle lever and killed the engine. The combine's virgin run had lasted exactly fifteen seconds.

"Hmmmm," Ted muttered. "I think we need to put in some more hydraulic oil."

Everyone was smiling. Mark jumped down from the cab. Alan asked Mark for his autograph. Alan was grinning lopsidedly, but he had made the request in earnest. He held out a pen and a slip of paper. As Mark signed it, Glen approached and grasped Mark by the shoulder. "I'd like to touch the man who started the first complete Bi-Rotor combine," Glen said. Mark's face, smudged with grease, broke into a weak smile.

"We'd better get it up on the stands," he said. Alan wheeled a big hydraulic floor jack into place under the combine's rear end. Dick Dinkins (a.k.a. Santa Claus) had brought the monster jack down from Peoria with one of his loads of goodies. The jack was about five feet long and had a large hydraulic cylinder in it. By harnessing a combination of levers and Pascal's principle, a person could pump a handle back and forth and lift twenty tons (or twenty-three and a half VWs) with the jack. Soon an inch of daylight showed between the concrete floor and the bottom of the rubber tracks.

10:00 a.m.: Joe and Ken fit the first skin panel on the back of the machine. Mark stepped back to appraise it. "That's looking a lot better already," he said.

11:40 a.m.: Alan and Ted finished topping off the hydraulic reservoir. "Time to rev it again," Alan said. To me, he explained, "We may have to fire it up for short bursts several times. You've got to fill all the hoses, then keep replenishing the reservoir."

Alan cleared the tracks of tools and debris. Mark climbed into the cab again. "Contact," he yelled. Ted gave the throttle a little jerk. Again the engine started immediately. Ted nudged the throttle a little, and the purr climbed to a higher pitch. Then one of the tracks started moving slowly around. Someone yelled, "Hey, the track's spinning!" Ted, who had been checking gauges, lunged for the throttle. In his haste, he pulled the lever down instead of pushing it up. The engine revved and the track spun faster. Then there was a sound like a giant champagne cork popping. Hydraulic oil was spewing everywhere. Ted killed the engine, and the gusher subsided to a hemorrhaging gurgle. Alan emerged from under the engine platform, drenched in a warm, sticky liquid. The stuff looked like Karo syrup but smelled strongly of petroleum.

"Shit," Alan said, looking at himself.

Jeff said, "You can say that again."

"Shit," Alan repeated. He wasn't hurt; just covered with oil.

"You've been slimed," someone said. Alan went away to clean up and change clothes.

Ted looked things over for a moment, then rendered his diagnosis. "One of the temporary filters blew. It's not designed to take much pressure, and when the engine revved it blew." The eruption could be rectified by replacing the filter, but the spinning track was a deeper and more vexing problem. “The transmission was in neutral, wasn't it?” Ted asked Mark. Mark confirmed that it was. The track shouldn't have budged. Ted and Mark began trying to figure out what was wrong.

1:30 p.m.: James Taylor's green-blue Cadillac pulled into the shop. James, Bernard, and a third man got out. Bernard introduced the dignified stranger as a vice president from Learjet. The new arrivals looked over the combine admiringly. “I'm sure proud of what these guys have done,” Bernard told his guest. “You look at it close, and it doesn't look like a typical cobbled-together prototype. These guys are meticulous craftsmen.” The guy from Learjet nodded. He seemed to be impressed.

2:05 p.m.: Mark honked the combine horn and turned the key for the third time. There were no eruptions, but the errant track moved slowly again. It was the left track, moving clockwise at the rate of a few inches per second—a slow reverse. Ted cut the engine after thirty seconds.

Bernard stood near Jeff Hawkins. “I'm sure proud of what you guys have done,” he said.

“Yeah, looking at it kinda makes you feel good,” Jeff replied. “Today Mark started it for the first time. I've been waiting for this day for six or seven months.”

3:15 p.m.: The skin crew welded on a second big side panel, and the machine's finished lines begin to suggest themselves. For the second time that day most of the crew took out cameras. Flashbulbs popped. Ralph was among the photographers. He had smiled that day more than I'd seen him smile in quite a while. “There she is,” he beamed. “She looks just like the drawings.”

3:55 p.m.: A white Chrysler drove into the building. I was standing near the door it entered and could see that the car contained two men dressed in gray suits. I didn't recognize them. “Can I help you?” I asked.

“Where's Gordon-Piatt?” one of the men asked.

“It's up the road,” I said, pointing back toward the door they had entered. But they didn't stop. The car drove slowly toward the combine.

By now Mark had seen the car. He was striding rapidly toward it. "What do you guys want?" he demanded.

"We just want to know what Gordon-Piatt does here," answered one of the men. They were looking at the combine intently. Mark walked around to the rear of the car to look at the license plate. I think he expected it to say "John Deere." But it didn't; the car had standard Kansas plates.

Ralph talked to the men, asking what they wanted. They didn't say much. They just looked around. Their gazes returned frequently to the combine. Mark didn't like this one bit. He was getting more upset by the second. "What are they doing here?" he shouted to Ralph. The car never stopped. It rolled slowly toward the door at the far end of the building. Mark strode along beside the driver's window. "You get on out of here. This is private," he yelled. The car sped up a bit, and so did Mark. He was running beside it now. He slapped it hard on the side, shouting, "Go on, get out of here!"

The white Chrysler drove out the far door.

"They were probably just Gordon-Piatt's insurance guys, checking out the building," Ralph said. Mark was breathing heavily, staring at the door where the car had disappeared. Nothing would convince him that the men weren't industrial spies. He brooded silently for the rest of the day.

4:30 p.m.: A big thunderstorm hit. With the rain pattering on the roof, Alan and Ted puzzled over the problem of the moving track. I sat down on a crate and started talking with Sushil. Sushil had a gift for metaphor that I was only beginning to appreciate. "In my country," he said, "there is a wedding tradition of dressing up the bride. When she is all ready, a special van comes for her and takes her away. She goes and stays with her husband for seven days. Then she comes back in the van to be with her family again for a while."

I nodded idiotically, interested in the story but unsure yet of its point. Sushil could tell I didn't get it.

"It's the same with this combine," he explained. "Now we are busy dressing up the bride. Soon she will go away, but she will come back after seven days to be with us again."

Not long after that, Dick Dinkins pulled into the building in his semi rig, pulling an empty lowboy trailer.

Sushil looked at me and said, "The special van has arrived."

10:30 p.m.: Glen and his skin men fit the last side panel on the combine. When it was secure, Glen stepped back. "We skinned the Cat," he announced. Nerves were frayed, tempers shot, but once again the cameras came out and the bulbs popped.

"Damn, she looks good," Jeff said.

Thursday, August 12: Another sweltering day. After Wednesday's dramatic progress, this would be a frustrating day of detail work and troubleshooting. Ted and Alan could not pinpoint the hydraulic problem that was making the track move. If they didn't solve it by the following day, Ralph's offer of free beer would expire. Far worse than that, though, was the prospect of sending an immobile and malfunctioning combine to Peoria. Their suspicions now centered on one hydraulic pump. But Ted and Alan had repeatedly checked all the fittings going to and from the pump and could find nothing wrong.

Friday, August 13, 9:00 a.m.: The hydraulic system was still screwing up. The problem was definitely in the pump that powered the left tread. The tread would spin backward when it wasn't supposed to, and it refused to spin forward at all. Alan called the salesman from the supplier who had sold them the pump, in Wichita. The salesman, Darrell Lile, was an extremely good troubleshooter of hydraulic systems, and he knew his products inside and out. Lile said he would come as soon as he could, and arrived in midafternoon. He disassembled the pump and discovered that it contained the wrong gasket—a piece of thin cardboard that fit between the head and cylinder casing of the pump. This gasket had the wrong pattern of holes in it. The nearest replacement gasket was in Salina, a two-hour drive away. Lile drove to Salina to get a twenty-cent piece of cardboard that was causing a debilitating problem with the combine. The faulty gasket was a pea under the XBR2's mattress.

Ralph spent most of the day applying body putty to spot welds on the skins. He sanded the hardened putty with a disk sander, smoothing the dimples to perfection.

9:45 p.m.: Lile returned from Salina. He hustled from his car to the combine carrying the new gasket carefully, as if he held a donor organ for transplant. The operation took twenty minutes.

10:10 p.m.: Alan climbed into the patient's cab and started its engine. Everyone watched the left track anxiously. "Go forward, you sonofabitch," said Aaron. His uncle, Alan, shifted the combine into forward gear. Both tracks started moving in unison, counterclockwise—forward. Victory whoops echoed throughout the building.

10:30 p.m.: The cameras came out again, as did a case of beer. Much preparation remained to be done before the painter could do his work. The cab, among other things, had to be masked with paper; the whole machine had to be rubbed down with methyl ethyl ketone, a powerful solvent, to remove oil and dirt; and a few more electrical and mechanical details remained to be finished up. It was clear, however, that the crew had met Ralph's challenge. The next night, there would be a party.

Saturday, August 14, 11:30 a.m.: Mike Logan, the painter, pulled into the shop in his pickup and unloaded his paint pot, spray gun, hoses, and other gear. He was soon ready to paint, but the combine crew would not relinquish the machine. Logan stood aside and watched the crew buzz around the combine, attending to a hundred last-minute details.

6:30 p.m.: The combine was finally masked and rubbed down. "It looks like it could be midnight before we get this thing shot," Logan said. He wore a white Tyvek "spacesuit" and was testing his painting equipment.

"Hell, no," Ralph replied. "We've gotta be done by eight. We're goin' dancin' tonight." He did a little two-step for emphasis. "Let's just get her primed now. The prime coat has to dry overnight, anyhow. We'll finish shooting her tomorrow. This is a going-away party for our baby."

Throughout the afternoon, family and friends had been arriving to join the party that night. Glen's wife, Patti, and their two children drove up from Texas with Glen's mother. The Jacksons were marking the combine's send-off with a family get-together. They had booked rooms at a motel in Ark City and hired a baby-sitter for the night. Dawn, Ralph's fiancée, made the two-hour drive down from the Pizza Hut where she worked, and arrived in the early evening. She was an energetic woman in her twenties with brunet hair, a pert nose, and jeans that flattered her trim figure.

9:30 p.m.: I rode with Alan down to Norm's Country Club, a popular night spot in Oklahoma about twenty miles south of Ark City. When we arrived, we soon spotted Jeff, Joe, and Aaron, who had staked out several tables at one end of the dance floor. The advance team had already procured thirteen pitchers of beer, which stood ready on the tables. "This is just for starters," said Jeff, beaming.

Norm's main room was the size of a roller rink, much of it filled with men in cowboy hats and women in western jeans. A country-rock band plucked and fiddled on a wide stage, while the dance floor—itself as large as a basketball court—pulsed with the counterclockwise Brownian motion of two-stepping couples. One corner of the vast space was fenced off to form a roping corral. For a dollar, you could mount a full-size plastic horse and have three chances to rope a mechanical calf, which an attendant shot out of a spring-loaded gate.

The Bi-Rotor tables were soon a merry beehive, with couples constantly coming and going to dance or get more beer. Mike Elswick was there with his wife. Mike Logan, the painter, was there, and so was Ted Schwitters, the plumber from Peoria. Around midnight someone said, "Hey, where's Mark?" Nobody had seen him. He hadn't come.

Sunday, August 15, 10:00 a.m.: Mike Logan finished shooting black trim on the combine's undercarriage, then began spraying its upper body bright Cat yellow. Everyone was there to watch except Mark, who didn't show up at the shop all day.

That evening I asked him where he had been. "I did a little fishin'," he said. "It's like when we were working on Whitey down in Texas. I got so wound up that I had to do something to relax a little. I told Ralph one day I was going to Vegas for the weekend. Ralph just said, 'All right, do what you gotta do. Just be careful.' So I drove to the Dallas airport, got onto a plane, and went. I played some craps, got me a nice room, had a good time. When I came back after a couple days, I felt refreshed and full of new ideas."

He lit a cigarette and took a long drag. "That's the point I'm at now," he added. "I figured I needed to go dunk a fishing line in the water. You know, when Edison was trying to figure out how to make an incandescent bulb, he was feeling frustrated, so he went out fishing. He cast his line out at night and the line glowed in the moonlight. That's when he had the idea of using a monofilament in the bulb. I need to do some fishing."

I asked him if he felt bad about sending the machine away.

He nodded. "I don't want to make a mistake at the last minute. It's like ol' Preston Tucker. He let that board of directors get involved, and it ended up costing him the whole project. You've got to watch out every step of the way. I don't like seeing the machine get away from us like this."

Monday, August 16, 6:15 a.m.: Mark and Ralph arrived at the shop as Dick Dinkins checked the chains that secured the XBR2 to the flatbed trailer. "We're here to see our baby off," Ralph said jubilantly. Mark walked slowly around the flatbed trailer, looking over his machine for the first time since it had been painted. The shiny yellow body seemed to glow softly in the early-morning light.

"It doesn't look like a prototype, does it?" he said finally. "It looks like our first production machine."

"Those guys at Caterpillar are going to be shocked," Ralph said. "They weren't expecting this. They were here a few weeks ago and said, 'Wow, you guys sure have a lot to do.' They're expecting a bare-bones prototype, and we're delivering *this.* It looks awesome."

7:15 a.m.: The whole combine crew assembled around the semitrailer, along with Mike Elswick and Jim Salomon. A blaze-orange "Wide Load" sign and warning flags decorated the machine, fore and aft. The flags looked festive as they fluttered in a light breeze. Dick Dinkins climbed into the cab of the truck, gave a little burst on the air horn, and drove away with the bride.

THE WEDDING

The problem with marriage is that it ends every night after making love, and it must be rebuilt every morning before breakfast.

—Gabriel García Marquez (1985)

THE CATERPILLAR MEETING took place in a large hangar at an airport industrial park near Peoria. About 200 corporate personnel attended, all of them on a hush-hush, need-to-know basis. The whole thing was, in the main, a dog-and-pony show, a kind of corporate pep rally intended to fire up the dealers about some new products Caterpillar would soon be introducing. A handful of machines were scattered around the hangar floor. The largest of them was a new hydraulic excavator, with the elbow of its apelike arm sticking twenty feet into the air. There were two updated models of the company's big Challenger tractors. Parked in the middle of the display, like the jewel in the crown, was the prototype of a completely new, relatively small farm tractor called the Challenger 45. Cat's ag-machine managers were very excited about the new tractor. It would be the industry's first all-purpose, midsize farm tractor that ran on tracks. The Challenger 45 had sleek, aerodynamic lines inspired by recent trends in automotive styling, which made the smaller tractor look like a sports car surrounded by dump trucks. It was sure to cause a sensation forty days hence, when it would be unveiled publicly at a farm-equipment show in Iowa.

Standing over to one side, also in the hangar, stood the XBR2. A combine of a different color was parked next to the Bi-Rotor for comparison. The other machine was a John Deere Maximizer retrofitted with Cat tracks. The big John Deere stood out like a green bean on a plateful of sweet corn.

On the morning of August 19, busloads of eager Cat dealers arrived at the airfield. They streamed out of the buses and gathered near the hangar's closed doors. At the appointed moment, the doors swung open and the men (with a few women scattered among them) surged forward like kids on Christmas morning. The event's organizers expected the bulk of the crowd to gather around the prominently displayed Challenger 45 first. To their surprise, though, many of the curious dealers headed straight for the XBR2.

Ralph got a phone call from Peoria that evening. "They loved it," he said after hanging up. "Every indication is that the dealers were very enthusiastic. Now they want test results. We've got to get into the field as soon as possible to see how she performs."

Several days after the meeting in Peoria, Ralph heard a story, perhaps apocryphal, that the chairman of Caterpillar, Donald Fites, had recently run into an executive from another company at a trade show. During their conversation the other executive reportedly asked, "What are you guys doing with that combine down in Kansas?"

Fites replied that he didn't know. He had not yet been given any details about the Bi-Rotor project.

"Well," the other man supposedly said, "I hear there are a couple of loonies loose down there with a combine they're trying to develop. They'd just wind up costing you a lot of money."

THE LOONIES IN Kansas now faced yet another deadline, which seemed more pressing than ever: getting the combine into the field for the autumn harvest. Even though the XBR2 had made a good showing in Peoria, Cat wasn't ready to buy the rights. Mark, Ralph, and the team had yet to demonstrate if, and how well, their pretty prototype would actually work. Cat wasn't about to acquire an unproven machine, but until the machine proved itself, the company was paying between $20,000 and $30,000 a month to keep the project alive. Cat wanted to fish or cut bait as soon as possible, and it transmitted this sense of urgency to the crew in Kansas. Bob Petersen even came down to help out during the final weeks of preparation. At Caterpillar the combine was his baby, and he was anxious to see that no time was wasted during labor and delivery.

The crew members had difficulty getting themselves moving again after the big push for Peoria. "It was like we had won the Superbowl—the feeling of accomplishment when we sent it up to the meeting," Ralph said. "Man, it looked so good. But now we're waking up the morning after the big game only to find out—oh, my God—that was just the damn *semifinals.* It's hard to get yourself back up to the same energy level, even though you know the biggest challenge is still ahead."

The machine they had sent to Peoria was like a Hollywood set—all outer form and little inner function. Now they had to make the guts work. That would prove at least as challenging as finishing the self-propelled shell had been. The work that remained would not produce visually dramatic, and therefore maximally gratifying, results. What they faced now were details that came to be known in the north building as "little shit"—connecting cables, modifying skin panels to improve engine ventilation, truing the rotor, finishing the unloader, installing sensors, adding braces to the grain bin. The list went on and on.

"What did you do today, Jeff?"

"Oh, just a bunch more little shit."

The team received a blow a few weeks later, partway through September, when Glen left. He had accepted a position as a senior design engineer at Peterbilt. He'd known since he first came to work on the combine that this moment would come. Designing and building the Bi-Rotor prototype was a two- or three-month project that had stretched to six months and counting. Glen and his young family had grown weary of living apart. Then, too, the continued security of Glen's job (and everyone else's) with Mark and Ralph was by no means certain. Peterbilt, by contrast, was a large, stable company that offered a good salary and benefits. He had a responsible job, he would only have to work eight hours a day, and he could devote his remaining time to Patti and the kids. As a father, husband, and breadwinner, he felt he would be utterly irresponsible, *crazy,* to pass up the opportunity at Peterbilt. But the adventurer, the strong team leader, and even the engineer in him hated like anything to leave the XBR2 and the boys camped out at Gordon-Piatt. He was Peter Pan's chief lieutenant in Neverland; but the time had come to leave the lost boys, go home, and accept his responsibilities as a grown-up.

The boys had a going-away party for Glen at the Mule Barn. They pulled three bar tables together at one end of the dark dining room and ordered large quantities of beer and steak. The group was unusually subdued, drinking its beer and watching the nearby dancing, pool, and shuffleboard mostly in silence. Alan had brought in his special mug, though under the circumstances its "party-time" message seemed a little forced. Alan was busy writing something on a piece of paper. Asked what he was doing, he replied that his wife's birthday was the next day and he had forgotten until then to write a note on a card for her.

The jukebox played several songs before Alan finally stopped scribbling. Then he clanked his plastic mug with a steak knife and said, "Let's have a toast." Everybody looked at him and reached for a glass. Everybody except Ted Schwitters, who picked up a thick piece of Texas toast from his plate and raised it aloft, producing a sudden and general outburst of laughter. Now that the ice had been broken, Alan stood up and read from the scrap of paper he had huddled over. Talking loudly to be heard over Billy Ray Cyrus and the Friday-night Mule Barn crowd, he said, "I'd like to dedicate a little poem to Glen Jackson and the Bi-Rotor combine." This is what he recited:

There once were some men who built the Bi-Rotor.
It had Caterpillar tracks and a Caterpillar motor.

It took a lot of sweat and an occasional yell,
But when we fired it up it ran pretty damn well.

Long ago I asked Ralph, "Are these men properly skilled
To put together what Mark Underwood wants to build?"

He laughed and smiled and said, "Yes, indeed!
Just a bunch of Kansas loonies is all we need.

"We can design and build in a month or two
What it would take John Deere several years to do."

Then I thought to myself as Glen Jackson headed away,
What could we do to convince him to stay?

The next morning, outside the north building, Glen ran a gauntlet of hugs and slaps on the back. Then he got into his pickup and drove off. As the crew watched him go, there wasn't a dry eye in the parking lot.

· · ·

In spite of Glen's departure, Mark was feeling better. His baby had come back from its trip, and there had been no disaster. Part of him still expected the worst to happen any moment. Even so, the dark cloud that had enveloped his psyche had now lifted noticeably. One morning he walked up to me, clapped his hand on my shoulder with a broad grin, and said, "How ya doin', ol' buddy?" It was a small gesture, but it revealed a glimpse of the old Mark, the plucky guru who had been missing for much of the summer.

The reason for this uplift, no doubt, was the approach of the crowning moment, the moment the entire team had dreamed of for months, when the XBR2 would make its maiden run in a field of ripened grain. A palpable sense of excitement spread in the north building as the long list of little shit grew shorter and shorter.

I had gone home for a while, but on September 14, Ralph called to summon me back to Kansas. "We'll finish the little shit on Friday, the twenty-fourth," he said. "That weekend, we'll go to the field."

I arrived in Winfield on the evening of September 24. The optimist in me expected to see the combine loaded on a trailer, greased and ready for the trip to a nearby farmer's field the next morning. Another part of me—pessimist? pragmatist? skeptic?—expected to see the combine's guts spilled all over the shop floor, not even remotely ready for a test run.

The second scenario proved correct. From the moment I walked into the shop, I knew the big moment would not take place the next day, or even the day after that. The Bi-Rotor assembly, which Ralph had personally spray-painted a flashy red, yellow, and silver, lay out in the open. It was attached to the combine's feeder house, which was in turn affixed to a thirty-foot-wide grain head. The whole mishmash lay sprawled in front of the combine's gaping mouth, as though a cat had just lost its supper. And it was no wonder: The cat's rear end seemed to be in pretty tough shape, too. The skin panels around the engine had been removed, and Ted Schwitters was back there messing with the abdominal plumbing again. Bob Petersen, Ted's boss, was helping Ted. "Bosses are like diapers," Ted told me with a grin as Bob stood nearby. "They're all over your butt, and most of the time they're full of shit." Bob tried his best to look amused.

Bob had not, on the whole, felt very amused lately. Like me, he had come down to Gordon-Piatt expecting to see the machine ready for the field and had been disappointed to find how much work remained to be done. He chipped in to help, assuming the double role of Ted's assistant and the team's self-appointed manager. Each morning he called a meeting in Ralph's cramped office to run down the list of details to be addressed that day. He wrote the list on a white marker board that hung on the wall, checking each item off as it was accomplished. A few days after my arrival on the 24th, for instance, here is what the board said:

- air cond. switch
- bin reinforce
- rotating grate
- engine compartment baffles
- rotor: trim vanes, weld nuts
- trailer: hydraulic lines, lights
- sieves
- grain sump door
- header auger—feeder house shaft
- cab door
- oil-cooler shroud
- engine speed sensor
- power to feeder-house reverser
- unloader: discharge boot, hydraulic hoses
- align track
- park break

During some meetings, the list grew longer instead of shorter. One morning Alan added these items while Bob talked:

- chaff-discharge pan
- grain-loss monitor
- grain tank-cover tarp

Bob viewed most of the items on the list as legitimate and necessary; some, though, he thought were dispensable. He often felt that he was dealing with a bunch of imaginative perfectionists who lacked the seasoned prototyper's sense of when to leave well enough alone. "These guys want to reinvent everything," he told me one day. "That's okay to a point, but there comes a time when you need to get into the field and try it out. After we get some test results, they can come back into the shop and invent and refine all they want. But now we need to see how the basic machine works."

Mark looked at things from a different point of view. "We have to get it right," he said in the Winnebago late one night, sipping coffee. "I'm not going out half-cocked. I've done enough prototyping to know that you've got to take the time to do it right the first time. You do it right, and it will work."

A week and three days after I arrived, the combine still sat in the shop, immobile. The list on the office wall had seventeen unfinished items on it. The date was October 4—Monday morning. At 9:00 a.m., Bob called a meeting and everyone filed wearily into the office. He went down the list with his marking pen, discussing each detail with the team members who were working on it. Finally he said, "So how long will it take us to finish this list?"

Everybody fidgeted uncomfortably and stared at the floor. The truth, as they all knew, was that it would take days to finish the list.

"Okay," Bob continued, "let's prioritize these. Are there any that aren't absolutely necessary to get done before we go to the field?"

Bob ran down the list again, putting brackets around items that everyone agreed were not totally essential. Four items went into brackets.

There were a few moments of silence. Then: "Can we get into the field tomorrow?" Bob asked.

Mark, who sat in a swivel chair with his feet on the desk, spoke up. "My feeling is that we need to nail down some of these things today, then run the machine for four or five hours tomorrow, just to make sure nothing overheats or anything. Then we go to the field first thing Wednesday."

Bob nodded. "Okay. Wednesday."

Wednesday, October 6th, dawned cloudless and unseasonably hot. "This is wheat weather, not milo weather," Jeff Hawkins said as he showed up for work that morning, already warm in his jeans and sleeveless T-shirt.

"It's a good day for cutting," said Mark, addressing himself to the combine. A light, drying south breeze wafted up from the Gulf of Mexico. The day's predicted high was ninety degrees.

"Supposed to rain tomorrow," Jeff said. "If we don't get out today, we won't get another chance till it dries up again. Who knows when that'll be."

The whole crew was gunning for the field. A problem, however, had cropped up the night before, possibly throwing a wrench in the day's plans. At 9:00 p.m. Tuesday, the crew had used a farmer's portable auger to fill the combine bin with wheat from Mark's grain truck. The object was to test the combine's unloading system, and the results had been discouraging; grain trickled out the end of the unloader when it should have gushed. They traced the problem back to the hydraulic pump that operated the unloader—a used pump they had salvaged from the old Cat Challenger. For some reason the pump wasn't performing. Another pea had lodged somewhere under the princess's stack of mattresses. Now, before going to the field, they had to find that pea.

By noon on Wednesday they still hadn't found it. Ted Schwitters sat in the office poring over a technical manual for the malfunctioning pump and discussing the problem with hydraulic experts at Cat headquarters.

Ralph paced the shop floor, feeling helpless. "That's one thing about doing this on a low budget," he said. "We salvaged that pump off the Challenger to save a few hundred dollars. Now it's keeping us from going into the field."

Bob stood off to one side, watching the team at work with a bemused look on his face. "I hope we go to the field today," he said, "but there's an awful lot of hammering, sawing, and hydraulic testing going on for a field day. I don't know."

More problems surfaced that afternoon. The trailer that would be used to transport the combine to the field was acting balky. Once its brakes were activated they refused to release, causing the wheels to lock. Also, no one had yet installed lights on the trailer to make it street-legal—a piece of little shit that had been overlooked until now.

Finally, at 4:30 p.m., Mark voiced what everyone else was beginning to suspect. "It's getting too late in the day to go out now," he said. "Let's finish what we've got to finish, load her up tonight, and cut in the morning."

Rain clouds started rolling in sometime during the night. The next morning the clouds formed a low, gray ceiling over the prairie landscape. Wednesday's breeze had died, and in the stillness the humidity felt oppressive. The team gathered in the shop at 8:30 and began preparing to go to the field. They went about their various tasks with quiet deliberation, as if preparing for battle. The XBR2 was loaded onto its trailer, which was hitched behind Mark's grain truck. The trailer appeared to be riding low, so Mark welded some steel skid plates under the front corners in case the trailer scraped the ground. Ralph and Joe hitched a second, smaller trailer to Ralph's new Jeep Grand Cherokee and loaded the combine header onto it. As it sat on the trailer, the header was taller and nearly three times longer than Ralph's Jeep. "It looks like a bug pulling an elephant," he observed.

Jim Salomon arrived with two men in his car. Salomon introduced the two visitors as bankers. They wanted to see the prototype run. So did Mike Elswick, who rode in a separate car. The Taylors' Caddy also pulled into the shop shortly after 9:00 a.m., with Bernard and James inside. An hour later everything was ready. A long procession of vehicles formed in the parking lot, with Mark in the lead, driving the grain truck, the big yellow combine in tow. The procession headed south, to a field of milo owned by a farmer with whom Ralph had made arrangements. As the caravan passed by the Gordon-Piatt plant, a dozen workers came outside, waving and jabbing their thumbs into the air.

The appointed field lay only about two miles away, on a secluded gravel road. A farmer tilling a nearby field almost drove his tractor into a ditch when the combine and its motorcade appeared before him like a weird mirage. The crew set the hydraulic jack under the tongue of the combine trailer, then unhitched the trailer from the truck. As they did so the trailer, with the combine still on it, began rolling forward on the loose gravel. Its tongue fell to the ground with a resounding *kaboom.* All eyes focused on the combine. No damage had apparently been done. Mark quickly jumped up into the cab, started the engine, and eased the machine off the trailer. When he was clear of all obstacles, he spun the combine around, performing a neat 180-degree turnabout on the narrow road. "I'd like to see a John Deere do that," Sushil said gleefully.

Hitching the combine to the header took a while. The rain clouds began to look more threatening and a cool wind picked up. Finally, ten minutes before noon, Mark turned the entire machine, combine and header, toward the rippling milo. He throttled up and activated the rotor and cage, the sieves and elevator, the header and feeder house. The whole machine throbbed with noise and power. Mark lowered the head and hit the grain like a bull attacking a matador's cape. He drove twenty yards and stopped. The crew had been running along behind, and could tell that crop material was balling up and clogging the feeder house.

"It's wet, slimy stuff," Mark said as he jumped down from the cab. With a few wrenches, he tightened some clutch plates in the feeder house and pulled damp masses of milo heads and stems out of an access panel. Then he climbed back into the cab and revved the engine again.

"Now work, you sonofabitch," Ralph said nervously.

This time it did. Mark drove a quarter of a mile to the end of the field, spun around and drove back, combining milo all the while. The debris that came out of the rear end of the combine contained few seeds, and the bin contained almost no M.O.G. This was an especially impressive accomplishment in wet, small-seeded milo. Sushil was jubilant. "I knew this would be a great machine," he said.

The group of observers broke into applause, whistles, and hoots. Even the two buttoned-down bankers seemed caught up in the excitement. Bob stepped up and asked Mark if he could ride for a round or two. As he and Mark headed back down the field, Jeff walked over to Ralph. The burly welder was almost overcome with emotion. "Congratulations," he told Ralph, offering his hand.

"What do you mean, congratulating me?" Ralph yelled. "This is *our* machine. Congratulations to you, too! Look at that sonofabitch go!"

Ted, also watching, said, "This is pretty unusual—having a prototype perform this well on its very first run."

Ralph grinned. "That's the flip side of Mark's obsession with getting everything right before coming to the field," he said. "We get out here, and it works the first time."

Mark made eight rounds of the field and gave rides to almost as many people. Then, to general consent, he decided to call it quits for the day. The team had just finished loading the machine onto the trailer when the rain started. (It didn't let up for four days.)

The Bi-Rotor team. On ground, from left: *Mark Underwood, Jeff Hawkins, Ken Jackson, Sushil Dwyer;* on ladder, from top left: *Ralph Lagergren, Joe Lutgen, Alan Van Nahmen, Glen Jackson. (Mark Underwood and Ralph Lagergren)*

"Well," said Bob after everyone had returned to the shop, "it performed as advertised."

Ralph stood in white tie and tails at one end of the living room, waiting. Chris LeDoux, a national bull-riding champion turned country troubadour, sang "Look at You, Girl" from the speakers of a nearby CD player. It made an unorthodox wedding tune, but it was Ralph's favorite song, and seemed appropriate for the occasion.

You mean everything to me
And I'd do anything
To have you stay forever.
I'm an ordinary man
But I feel like I could do anything in the world
When I look at you, girl.

The bride descended the candle-lined stairway at the opposite end of the living room of her mother and stepfather's house in Little River, Kansas. About forty guests crowded around, leaving a narrow pathway for the bride. The ceremony was brief, conducted by a portly judge from Wichita. Following the vows, the bride's brother played another love ballad on the CD machine, this one sung by George Strait.

I'll cross my heart, and promise to
Give all I've got to give
To make all your dreams come true.

Everybody piled into cars following the ceremony and drove to an equipment shed on the edge of town. Ralph threw open the doors, and there was the combine, almost as radiant in yellow paint as the bride was in white satin. "Dawn thought she married a man," Ralph said, flushing red, "but she also married a thing." The wedding guests laughed. One woman asked, "Which comes first?"

Ralph wanted to carry Dawn across the threshold of the XBR2's cab but rejected the idea after reconsidering the steepness of the ladder and the length of his bride's wedding train. Instead, the newlyweds posed for pictures on the ground, in front of the Cat tracks. Then Ralph gathered the XBR2 crew together for more pictures. As they huddled for the cameras, Marsha Van Nahmen and Patti Jackson (the wives of Alan and Glen, respectively) chatted with Dawn. "Welcome to the combine widows' club," Marsha said.

The combine made a fine wedding prop, but it was only passing through Little River on business. It was on its way up to Burr Oak, to make its corn-cutting debut on Mark's parents' farm. Mark stood outside the equipment shed, smoking pensively while the guests admired his machine. He was eager to see how it performed in corn, and so was Caterpillar. The company was sending some dignitaries out during the coming week to watch the corn trials. Mark was confident that the machine would, once again, perform as advertised. What he was look-

Closing the circle: Mark cutting corn with the XBR2 on his home place. (Mark Underwood and Ralph Lagergren)

ing forward to most of all was simply driving the machine he had invented on his own home ground. For him it would be a vindication of his talents, the closing of a circle. "It'll be a good feeling," he said, "to take her out on the farm where I growed up."

There was, of course, a dance that night. It took place at the American Legion post in Lyons, the nearby county seat. More than 150 guests filled the hall to overflowing. A tuxedoed DJ led the exuberant group through the Grand March, the Chicken Dance, the Flying Dutchman, and the Hokey Pokey, with plenty of country-western songs in between. The Bi-Rotor team occupied a place of honor by the beer keg, which sat beside the stage where the DJ worked. The team members considered it their personal responsibility to see that all five kegs were duly consumed. Mark was particularly conscientious in carrying out this duty. By 2:00 a.m., when the music stopped, he was looped.

"I've got one favor to ask," Ralph told Mark as the party broke up. Ralph and Dawn were about to leave for the Wichita airport; they planned to catch a 6:00 a.m. flight to Cancún for their honeymoon. "I want you and the guys to have the cake," Ralph said. "You guys take it up home and eat it there, because we're not taking it to Mexico."

In addition to the regular wedding cake, Dawn's mother had decorated a special cake with a remarkably detailed image of the XBR2 combine. She had traced the machine with pencil-thin lines of brown frosting. The tracks were black, and the body had been airbrushed Cat yellow. This was the cake Ralph was entrusting to Mark, but he was in no shape to carry it. I was slightly steadier on my feet, so I picked up the combine cake and accompanied Mark and Sushil out into the cool October night.

The next morning at 8:30, Mark stood out in the motel parking lot, staring into the open trunk of his car. He was holding the special wedding cake, searching for a place in the crowded trunk where it could ride home safely.

"How are you feeling, Mark?" I asked.

"Oh, fine," he said. "I had a bit of a thumper when I woke up, but it's gone now. Hey, do you have something I could put this cake in, to kind of protect it?"

I happened to have an empty shopping bag in my car, and went to get it. I held it open while Mark carefully put the cake inside.

He took the parcel and smiled.

"The XBR2 is in the bag," he said.

EPILOGUE

It ain't over till it's over.

—Yogi Berra (1973)

THE FIRST SIGNS of trouble with Caterpillar appeared a few months after Ralph returned from his honeymoon. Cat called to say that it needed help in evaluating the Bi-Rotor combine and was, therefore, enlisting the help of another firm. This second firm, which I will call "Company X," was a major player worldwide in the farm-equipment industry. The up side of its involvement with the Bi-Rotor, as far as Mark and Ralph were concerned, was that if the evaluations went well, Company X might develop and market the Bi-Rotor for the international market, while Caterpillar would market it in North America. The down side was that Company X, which manufactured its own line of combines, might have ulterior motives in the matter. X's people might be thinking: If Cat wants to put its name and tracks on a combine, let it be one of *our* combines. Mark, in particular, couldn't help wondering how impartial Company X would be in carrying out its investigations on his baby.

Company X maintained a research facility in Europe, and it wanted to test Mark's basic threshing concept there. The Bi-Rotor team was urged to build a new rotor and cage, similar to the threshing unit inside the XBR2. They were to ship the new unit to Europe, where it could serve as a mechanical lab rat.

The team built the unit, and Mark and Sushil accompanied it to the testing lab. Mark soon returned to Kansas to oversee modifications of the XBR2, but Sushil stayed behind for a month to monitor the testing.

Everything seemed to go well. The tests suggested a few minor adjustments that would improve the Bi-Rotor's threshing efficiency in certain crops and conditions. But the overall report was rosy.

That winter, Glen Jackson, who had reluctantly left the Bi-Rotor project to take a steady job at Peterbilt, stayed in close touch with Ralph and the team.

"You know," he told Ralph one day, "if you guys make it big without me, I'll be thrilled. But if you don't make it and I'm not there, I'll always kick myself for not hanging in with you. I'll always wonder what might have happened."

"Glen," Ralph replied, "I won't say we don't miss you. You were the glue that held the team together."

A few weeks later, Glen was back with the team."I'm sorry, but I made a mistake," he said when he announced his resignation at Peterbilt. "I started something up in Kansas that I've got to finish." His employers were sorry to see him go, but were sympathetic.

The following spring, Cat and Company X suggested another round of testing. They wanted to pit the XBR2 against two other machines in some of the most demanding harvesting conditions in the world—the wheat fields of the Imperial Valley. Essentially a fertile desert, this low trough in extreme southeastern California is intensely irrigated with water diverted from the Colorado River. As a result, the desert blooms with cotton, citrus, dates, and winter vegetables. Wheat grown in the Imperial Valley yields a phenomenal 100 to 130 bushels to the acre (compared with around 40 bushels per acre in a good, unirrigated Kansas field). For five grueling weeks, the Bi-Rotor team performed grain-loss tests between El Centro, California, and Yuma, Arizona. Temperatures there sometimes exceeded 110 degrees; a wrench left in the sun would burn the hands.

The other two combines being tested were commercial models, graduates of multimillion-dollar corporate R&D programs; the XBR2, by comparison, was a kid straight off the farm. In the trials the Bi-Rotor held its own, consistently rating second in performance among the three. Still, the heavy desert wheat did reveal a few weaknesses in its design. The problems were remediable, yet in the intense heat, sun, and isolation of the Imperial Valley they seemed glaring—especially with personnel from Company X looking on.

The machine that consistently came out on top during these head-to-head trials just happened to be one of Company X's models.

Mark, Ralph, and the other Bi-Rotor boys wondered if they had been set up for failure.

Not long after they returned to Kansas, Bob Petersen drove down from Cat headquarters with bad news: His company was going to stay out of the combine business for the time being. Caterpillar, he said, would be providing no further financial support. It was giving up its right of first refusal on the Bi-Rotor.

"We got caught with our drawers down," Ralph said, as he and the rest of the team absorbed this shock. "Now we've got to pull our drawers back up, tighten our belts a few notches, and tough it out." Some of the team members took part-time jobs in order to pay the rent. But nobody abandoned the project. The team went into suspended animation while Ralph scouted for funding.

For Mark, the spring of 1994 was especially difficult. Not only did the near-marriage with Cat fall apart, his own marriage did, too. Mark helped Deb buy the bar and grill in Burr Oak, where she found an appreciative audience for her fine home cooking. Mark, who had spent so much of his life repairing machines, now needed some time to repair his own spirits. And yet, after the setback with Cat, he felt he had to prove all over again that he and his machine were not failures.

Late in the summer of 1994, Ralph put together a deal with a Kansas venture-capitalist that gave the team another lease on life. They came back together at Gordon-Piatt and worked feverishly to refine the XBR2 so it would be at its best for the fall harvest.

At about that time, another suitor turned up, one that hadn't expressed much interest in the Bi-Rotor before. The courtship took place entirely behind a veil of secrecy and nondisclosure agreements. During negotiations that lasted through the winter, Ralph was wary. "The one thing I refuse to do," he said early in 1995, "is to accept a buyout at a fire-sale price. We're either going to get a fair amount for ourselves and our investors, or we're going to get zero. I'm not afraid to get zero." He *was* afraid, however, that this latest suitor intended to buy the Bi-Rotor combine only to let it die. The company, he worried, might be less interested in developing and selling Bi-Rotor combines than in removing this nettlesome upstart—and potential competitor—from the harvesting field.

To counter such a flanking maneuver, Ralph stepped up his grassroots promotional efforts. The team displayed the XBR2 at trade shows, attracting a good deal of interest and attention among prospective buyers. Ralph produced a video about the machine and sent out several thousand copies to farmers, institutional investors, and various news organizations. He hoped to make the Bi-Rotor so well known, and the object of so much pent-up demand, that the big companies would stop playing cat-and-mouse.

Ralph was still absolutely confident that the Bi-Rotor would emerge from this purgatory in triumph. The only thing that remained uncertain in his mind was what color it would be painted when it finally made it to market.

I stood in awe at the durability of his optimism. But I had to ask myself, was it courage or lunacy?

These two human traits, it seems, are engraved on opposite sides of the same coin. A final verdict on the sanity of any daring enterprise depends on whether, when the coin is tossed, it lands heads or tails. Success retroactively justifies and redeems all the efforts leading up to it, while failure, however undeserved, inevitably imposes the stigma of folly.

No matter what happened, Mark and Ralph had done a remarkable thing. Few independent inventors in the twentieth century have brought a new machine as big and complex as a whole combine so close to commercial adoption. There are, of course, no consolation prizes in the inventing business. As Ralph said, he and Mark would either score big or score nothing at all. They had followed their dream; whether it would come true now seemed to hinge on a cosmic toss-up. The coin hung in the air, spinning.

It landed the following summer, a year after the Bi-Rotor team had received the devastating news from Caterpillar. Mark and Ralph's new suitor stepped out of the wings and made an offer to buy development rights to the new combine. The offer, Ralph felt, was too low even to qualify as serious. It would barely get the boys out of debt, let alone begin to compensate them for their years of risk and sacrifice. He turned it down.

The suitor came back with another, considerably higher, offer. The terms still fell far short of the team members' hopes, but it was a concrete offer of cash. The boys could take the money and be done with it—perhaps at least saving face, if not striking it rich. At the meeting where this offer was made, Ralph didn't hesitate. He looked at his lawyer and said, "Come on, Scott, let's go home. I thought they were ready to talk turkey." They got up and walked away from the table. Ralph had called the toss, and it appeared that he had lost.

In July, 1995, the suitor called again. The parties re-assembled at the bargaining table. This time, it was clear, there would be no more chances. As it turned out, no more were needed. A deal was struck—the session ended with hand-shakes all around. Agri-Technology, L.P., the partnership Mark and Ralph had formed twelve years before, sold its patents and development rights to the company Mark and Ralph had set out to beat: John Deere.

"We have jointly agreed with Deere not to divulge the details of the relationship, including the amount paid," Ralph told a reporter for *Soybean Digest.* "But we got a fair deal; we have been paid pretty well for all those years we had to tighten our belts and do without. Our financial partners at Agri-Technology also earned a good return on their investments."

Under the multimillion-dollar agreement with Deere, Mark and other team members would continue to test and refine the technology. Their ability to meet or exceed certain performance goals would determine the Bi-Rotor's ultimate fate. "John Deere has seen the potential in Mark's inventions," Ralph said. "Now we'll see if we can end up with the finest machine in the world."

ACKNOWLEDGMENTS

Ever since my family and I moved to my late grandparents' Iowa farm in 1988, I've wanted to write a book about the place. Two facts about "Canine Corner," as local old-timers know it, struck me as possible leads to an interesting story.

Fact 1: As recently as the 1940s, this small, diverse farm provided a living for a family of five.

Fact 2: By the 1980s (indeed, well before then), this little patch of the world's richest soil had become an agricultural fragment—the merest fraction of the thousand-or-so acres now required to support a family by farming.

What had happened between the 1940s and the 1980s to account for this transformation? (The critical period in agricultural history, I soon learned, was actually from 1840 to 1940; the past fifty years represent the extension of trends that were already riding on greased rails.) Whatever caused the change, it seemed to transcend Canine Corner, and even to transcend the subject of agriculture itself. After all, not many generations have passed since the day when the business of America was not business, but farming. The metamorphosis of my grandparents' small Iowa farm from an icon of America's agrarian legacy into a minor agribusiness asset seemed to me a change of large significance in the life of the nation. Somehow, I wanted to write about it.

The opportunity arrived one day with a call from my literary agent, Sallie Gouverneur. She told me that a well-known foundation based in New York planned to sponsor a series of books on the defining technologies of the twentieth century. Among the subjects the foundation hoped to include in its series was agriculture.

Suddenly I had a focus for my scattered thoughts on farming. I wrote a proposal, and this book is the result. My greatest debt of gratitude, therefore, is to the Alfred P. Sloan Foundation—especially Art Singer—and to Sallie Gouverneur, who served as the book's catalyst and midwife.

My next greatest debt of gratitude is to Mark Underwood and Ralph Lagergren, whose efforts to reinvent the combine and bring the invention to market became the narrative backbone of this book. Mark and Ralph entrusted me with their confidence and made me feel like a member of the Bi-Rotor team. The other members of the team, as they came on board, also accepted me with perfect friendliness and candor. I am profoundly grateful to them all.

Several other people were generous with their time and commendably tolerant of a writer who kept hanging around and asking questions at times when, no doubt, they wished he would go away. I'm thinking especially of people who work for various companies named in the book, particularly Caterpillar, Gordon-Piatt Energy Group, DuPont, Deere & Co., and Kincaid Equipment Manufacturing. I mention most of these people by name in the book. In a few cases, however, I was unable to use real names because of corporate policies and legal considerations. I regret this, because these people deserve credit for their courage, initiative, and hard work, not all of which were rewarded in the end. Identified or not, they all have my sincerest respect and thanks.

Many thanks are due, as well, to Tom Urban, Ken Madden, and Roger Parsons of Pioneer Hi-Bred; Dr. Donald Duvick, formerly of Pioneer Hi-Bred; Dr. Terry Sharrer of the Smithsonian Institution; Dr. Kurt Sundelin and Dr. Elmo Beyer of DuPont; Brad Agry and Russell Flinchum at Henry Dreyfuss Associates (the firm responsible for the industrial design of John Deere tractors and equipment since the 1920s); Raymond Frank and Norman Covert at Fort Detrick, Maryland; and, finally, my neighbor and farming mentor, Larry Ellis. All of these people were extremely generous with their time and expertise. Though none of them makes an explicit appearance in the book, they all made me better informed in important areas.

Eric Schlosser read an early draft and offered many insightful and detailed suggestions, all of which helped make this a better book. I hope I can return the favor someday. He and the others mentioned here all deserve credit for the good bits between these covers; any shortcomings are, of course, purely my own responsibility.

Virginia Wadsley did yeoman (or, I should say, yeowoman) work in obtaining permissions, credits, and prints for the illustrations. She's a gem of the prairie, and I thank her.

Bobbie Bristol, Jenny McPhee, and Elizabeth Sheinkman, at Knopf, also did yeowoman work in shepherding me and this book through the publishing process. Deepest thanks to them and to Sonny Mehta for having faith in me.

I'd also like to mention a few of the many teachers and editors who have guided, inspired, and instructed me as a writer. They include Emily Schaumburg at Cowles Elementary School (fourth grade), Tom Mathews at *Newsweek,* Tom Rawls at *Harrowsmith,* Barry Estabrook at *Eating Well,* and John McPhee, whose course at Princeton changed my life.

My aunt Marjorie and uncle Wayne were generous in sharing their memories of life at Canine Corner when there was a working, self-sufficient family farm here.

On the back of their first Bi-Rotor combine, Mark Underwood and Ralph Lagergren painted these words: "Dedicated to our parents, who dared to let us dream." Here, near the back of my first book, I'd like to express the same sentiment to my parents, Helen and Willard Canine.

Thanks transcending all others must go to my ultimate editor, colleague, friend, wife, and parenting partner, Molly Coxe. She made the words, and the work, flow better. More important, she and our two children, Will and Franny, reminded me often that there's more to life than words, words, words.

C. C.

SOURCES

Rather than dividing my list of principal sources by genre (books, articles, and unpublished materials) in the conventional manner, I thought it better to carve up the list along the lines of major subjects and themes. There's a section of references (I) about Mark Underwood, Ralph Lagergren, and the Bi-Rotor combine, followed by a section (II) that lists works of general agricultural history. After that come sections on each of the three major categories of agricultural technology discussed in this book: (III) mechanical technology (with the combine and tractor serving as my major examples), (IV) genetic technology (hybrid corn), and (V) chemical technology (the herbicide 2,4-D). Finally there's a section (VI) listing my major sources on general (not specifically agricultural) history, followed by a catch-all section (VII) listing references on important miscellaneous aspects of agriculture and agricultural history.

But first I'd like to acknowledge particular debts to a few books, since I had them open on my desk more often (and spilled more coffee on them) than the rest.

Graeme Quick and Wesley Buchele's book *The Grain Harvesters* was my most important source of information on the history of the combine.

The Romance of the Reaper, Herbert Casson's turn-of-the-century paean to Cyrus McCormick and his mechanical legacy, could hardly be considered unbiased history. Even so, Casson was a reporter as well as a booster, and personally interviewed nearly all the "reaper men" who appear in his lively little book. I made use of many of his quotations and anecdotes, but tried whenever possible to corroborate his version of events with accounts from other sources.

John Deere's Company is a corporate history of the exemplary kind. Its author, Wayne Broehl, is a careful researcher and a lucid writer who not only had complete access to company records but (though he's fairly uncritical) also had complete scholarly independence. His book puts the history of John Deere's company into a broader national and international context so that it ends up providing a good overview of American agricultural history from the 1830s to the 1980s.

I drew on William Cronon's book *Nature's Metropolis* for a number of its unsurpassed features: its magisterial notes and bibliography, its detailed account of the growth of the grain trade in Chicago and the Great West, and most of all, its inspiring example of historical scholarship coupled with a talent for interpreting the past and recounting it with narrative power.

I found myself referring to Thomas P. Hughes's book *American Genesis* over and over for its compelling portraits of great inventors. Hughes helped me to think about "my" farmer-inventor, Mark Underwood, in terms of the characteristics he shares with independent American inventors of the past. On a more general level, *American Genesis* helped me to gain a better conceptual grasp of what Hughes, in the book's introduction, calls "the history of modern technology and society in all its vital, messy complexity."

I

Virtually all of the material in this book about Mark, Ralph, and the Bi-Rotor combine comes from my own interviews and reporting. Many newspaper and magazine articles about the Kansas cousins have been written, however, and I have benefited from several of them—especially from the work of Larry Reichenberger of *Farm Journal,* whose September 1991 article first alerted me to the existence of Mark, Ralph, and their "dream reaper" (a fine headline that appeared on one of Reichenberger's subsequent *Farm Journal* articles about the Bi-Rotor boys).

Barnes, Anna. "Radical Rotary Combine." *Farm Industry News,* July–August 1991, p. 11.

Burkhead, Jeff. "Farmer Cultivates Field of Dreams." Farm Bureau *Farm Leader.* Vol. 9, no. 37 (Feb. 21, 1992), pp. 1ff.

Carlson, Eugene. "Farmers Reap New Business with Crop of Inventions." *Wall Street Journal,* January 29, 1992, p. B2.

"Class V Combine with Class VII Capacity." *Farm Industry News.* September 1992, p. 10.

Degnan, Joseph. "Big-ticket Items Built to Buck the Trends." *Farm Industry News,* January 1993, p. 79.

"Entrepreneurship: Development of the Bi-Rotor Combine Demonstrates the Value of Hard Work, Persistence and Free Market Risk-taking." *Arkansas City Traveler,* October 25, 1993, p. 4.

Ernst, Hank. "Harvesting a Dream." *Kansas Farmer,* January 1994, pp. 74–75.

Houtsma, Jim. "Bi-Rotor Still on Front Burner." *Farm Industry News,* February 1994, p. 46.

Painter, Steve. "On the Cutting Edge." *Wichita Eagle,* January 19, 1992, 1f.

———. "Innovative Combine Is on Fast Track." *Wichita Eagle,* October 31, 1993, 1F–2F.

Peterson, Chester, Jr. "Radical Bi-Rotor Wonder." *Corn Farmer,* spring 1992, n.p.

Reichenberger, Larry. "A combine with a Twist." *Farm Journal,* September 1991, p. 22.

———. " 'Bi-Rotor' Combine Whips up Interest." *Farm Journal,* October 1992, p. D2.

———. "Dream Reaper." *Farm Journal*, February 1994, p. B4.

Stowe, David. "Strother Firm Helps Build New Kind of Combine." *Winfield Daily Courier*, October 23, 1993, pp. 1, 6.

"Bi-Rotor Combine Enters Round Two." *Working Tires*, fall 1992, pp. 12–13.

Wedel, K. A. "Cousins Making Harvesting History." *Arkansas City Traveler*, October 22, 1993, pp. 1F, 5F.

II Works of general agricultural history

Benedict, Murray R. *Farm Policies of the United States, 1790–1950: A Study of Their Origins and Development*. New York: Twentieth Century Fund, 1953.

Bogue, Allan G. *From Prairie to Cornbelt: Farming on the Illinois and Iowa Prairies in the Nineteenth Century*. University of Chicago Press, 1963.

Ferleger, Lou, ed. *Agriculture and National Development: Views on the Nineteenth Century*. Ames: Iowa State University Press, 1990.

Heiser, Charles B., Jr. *Seed to Civilization: The Story of Food*. New ed. Cambridge, Mass.: Harvard, 1990.

Lord, Russell. *The Care of the Earth: A History of Husbandry*. New York: Thomas Nelson & Sons, 1962.

Rasmussen, Wayne D., ed. *Agriculture in the United States: A Documentary History*. 4 vols. New York: Random House, 1975.

Schlebecker, John T. *Whereby We Thrive: A History of American Farming, 1607–1972*. Ames: ISU Press, 1975.

Shannon, Fred A. *The Farmer's Last Frontier: Agriculture 1860–1897*. New York: Farrar & Rinehart, 1945.

III Mechanical farming technology

Broehl, Wayne G., Jr. *John Deere's Company: A History of Deere & Company and Its Times*. New York: Doubleday, 1984.

Casson, Herbert N. *The Romance of the Reaper*. New York: Doubleday, 1908.

Council for Agricultural Science and Technology. *Agricultural Mechanization: Physical and Societal Effects, and Implications for Policy Development*. Report no. 96. Ames: Iowa State University Press, February 1983.

Deere & Company. *Power Farming with Greater Profit*. Moline, Ill.: Deere & Co., 1937 (reprint of John Deere Centennial catalog).

"Deere, John." *Dictionary of American Biography*. New York: Charles Scribner's Sons, 1930.

Gray, R. B. *The Agricultural Tractor: 1855–1950*. St. Joseph, Mich.: American Society of Agricultural Engineers, 1975.

Greeno, Follett L., ed. *Obed Hussey: Who, of All Inventors, Made Bread Cheap*. N.p., Follett Greeno, 1912.

Higgins, F. Hal. "John M. Horner and the Development of the Combined Harvester." *Agricultural History,* XXXII no. 1 (1958), pp. 14–24.

Hurt, R. Douglas. *American Farm Tools: From Hand-Power to Steam-Power.* Manhattan, Kansas: Sunflower University Press, 1982.

Hutchinson, William T. *Cyrus Hall McCormick.* (2 vols.) Vol 1: *Seed-Time, 1809–1865.* New York: Century, 1930.

Leffingwell, Randy. *The American Farm Tractor: A History of the Classic Tractor.* Osceola, Wis.: Motorbooks International, 1991.

MacMillan, Don, and Roy Harrington. *John Deere Tractors and Equipment: Vol. 2, 1960–1990.* St. Joseph, Mich.: American Society of Agricultural Engineers, 1991.

MacMillan, Don, and Russell Jones. *John Deere Tractors and Equipment: Vol. 1, 1837–1959.* St. Joseph, Mich.: American Society of Agricultural Engineers, 1988.

Marsh, Barbara. *A Corporate Tragedy: The Agony of International Harvester Company.* Garden City, N.Y.: Doubleday, 1985.

McCormick, Cyrus. *The Century of the Reaper.* Boston: Houghton Mifflin, 1931.

McKinley, Marvin. *Wheels of Farm Progress.* St. Joseph, Mich.: American Society of Agricultural Engineers, 1980.

Quick, Graeme R., and Wesley F. Buchele. *The Grain Harvesters.* St. Joseph, Mich.: American Society of Agricultural Engineers, 1978.

Wik, Reynold M. *Henry Ford and Grass-roots America.* Ann Arbor: University of Michigan Press, 1972.

———. "Henry Ford's Science and Technology for Rural America." *Technology and Culture.* Vol. 3 (1962), pp. 247–58.

———. *Steam Power on the American Farm.* Philadelphia: University of Pennsylvania Press, 1953.

Williams, Robert C. *Fordson, Farmall, and Poppin' Johnny: A History of the Farm Tractor and Its Impact on America.* Urbana and Chicago: University of Illinois Press, 1987.

IV The development of hybrid corn

Crabb, A. Richard. *The Hybrid-Corn Makers: Prophets of Plenty.* New Brunswick: Rutgers University Press, 1948.

Duvick, Donald N. "The New Biology: A Union of Ecology and Molecular Biology," in *Choices: The Magazine of Food, Farm, and Resource Issues.* Vol. 5, no. 4 (1990), pp. 4–7.

Fitzgerald, Deborah. *The Business of Breeding: Hybrid Corn in Illinois, 1890–1940.* Ithaca: Cornell University Press, 1990.

Fussell, Betty. *The Story of Corn.* New York: Knopf, 1992.

Kirkendall, Richard S. *Uncle Henry: A Documentary Profile of the First Henry Wallace.*

Ames: Iowa State University Press, 1993.

Mangelsdorf, Paul C. *Corn: Its Origin, Evolution and Improvement.* Cambridge: Harvard Belknap Press, 1974.

Mosher, Martin L. *Early Iowa Corn Yield Tests and Related Later Programs.* Ames: Iowa State University Press, 1962.

Roe, Keith E. *Corncribs in History, Folklife, and Architecture.* Ames: Iowa State University Press, 1988.

Wallace, Henry A., and William L. Brown. *Corn and Its Early Fathers.* Revised ed. Ames: Iowa State University Press, 1988.

V Herbicides and the development of 2,4-D

Brophy, Leo P., and George J. B. Fisher. *The Chemical Warfare Service: Organizing for War.* Washington, D.C.: Department of the Army, 1959.

Brophy, Leo P., Wyndham D. Miles, and Rexmond C. Cochrane. *The Chemical Warfare Service: From Laboratory to Field.* Washington, D.C.: Department of the Army, 1959.

Carson, Rachel. *Silent Spring.* Greenwich, Conn.: Fawcett Publications, 1962.

Crockett, Lawrence J. *Wildly Successful Plants: A Handbook of North American Weeds.* New York: Collier Books, 1977.

"Ezra J. Kraus Dies" (obituary). *New York Times,* March 1, 1960.

Guerrero, Peter F. "Pesticides: 30 Years Since *Silent Spring*—Many Long-standing Concerns Remain." Statement before the U.S. House of Representatives Subcommittee on Environment, Energy, and Natural Resources, Committee on Government Operations. U.S. General Accounting Office, July 23, 1992.

Haber, L. F. *The Poisonous Cloud: Chemical Warfare in the First World War.* Oxford: Clarendon Press, 1986.

Hamner, Charles L., and H. B. Tukey. "The Herbicidal Action of 2,4-Dichlorophenoxyacetic and 2,4,5-Trichlorophenoxyacetic Acid on Bindweed." *Science,* vol. 100, no. 2590:154-155 (Aug. 18, 1944).

Hayes, Wayland J., Jr. *Toxicology of Pesticides.* Baltimore: Williams & Wilkins, 1975.

Hersh, Seymour M. *Chemical and Biological Weapons: America's Hidden Arsenal.* Garden City, New York: Anchor Books, 1969.

Holm, LeRoy G., Donald L. Plucknett, Juan V. Pancho, and James P. Herberger. *The World's Worst Weeds: Distribution and Biology.* Honolulu: University Press of Hawaii, 1977.

Janick, Jules, Carl H. Noller and Charles L. Rhykerd. "The Cycles of Plant and Animal Nutrition," In *Food and Agriculture,* A Scientific American Book. San Francisco: W. H. Freeman & Co., 1976.

Kleber, Brooks E., and Dale Birdsell. *The Chemical Warfare Service: Chemicals in Combat.* Washington, D.C.: Department of the Army, 1966.

Kraus, Ezra J., ed. "Studies on Plant Growth-Regulating Substances," in *Botanical Gazette,* Vol. 107 (June 1946), *passim.*

Marth, Paul C., and John W. Mitchell. "2,4-Dichlorophenoxyacetic Acid as a Differential Herbicide. *Botanical Gazette,* 106, pp. 224–32 (1944).

McCarthy, Richard D. *The Ultimate Folly: War by Pestilence, Asphyxiation, and Defoliation.* New York: Knopf, 1970.

Mitchell, John W. "Plant Growth Regulators," in *Science in Farming: USDA Yearbook of Agriculture 1943–1947.* Washington, D.C.: U.S. Government Printing Office, 1947.

National Research Council. *Pesticides in the Diets of Infants and Children.* Washington, D.C.: National Academy Press, 1993.

———. *Regulating Pesticides in Food: The Delaney Paradox.* Washington, D.C.: National Academy Press, 1987.

Peterson, Gale E. "The Development of Chemical Weed Killers in the United States." M.A. thesis, University of Maryland, 1968.

———. "The Discovery and Development of 2,4-D." *Agricultural History,* Vol. 41 (July 1967): pp. 243–53.

Pokorny, Robert. "Some Chlorophenoxyacetic Acids." *J. Am. Chem. Soc.* 63 (June 1941), 1768.

Stein, Sara B. *My Weeds: A Gardener's Botany.* New York: Harper & Row, 1988.

Thomson, W. T. *Agricultural Chemicals. Book II: Herbicides.* Fresno, Calif.: Thomson Publications, 1981.

U.S. Institute of Medicine. *Veterans and Agent Orange: Health Effects of Herbicides Used in Vietnam.* Washington, D.C.: National Academy Press, 1993.

"Weed Killers." *Kirk-Othmer Encyclopedia of Chemical Technology.* Second ed. New York: Wiley & Sons, 1970.

Weed Science Society of America. *Herbicide Handbook.* Sixth ed. Champaign, Ill.: Weed Science Society of America, 1989.

Whorton, James. *Before Silent Spring: Pesticides and Public Health in Pre-DDT America.* Princeton: Princeton University Press, 1974.

Wines, Richard A. *Fertilizer in America: From Waste Recycling to Resource Exploitation.* Philadelphia: Temple University Press, 1985.

VI General (nonagricultural) history, natural history, and the history of technology

Allen, Oliver E. "The Power of Patents." *American Heritage,* September–October 1990, pp. 47–59.

Boorstin, Daniel J. *The Republic of Technology.* New York: Harper & Row, 1978.

Braudel, Fernand. *The Structures of Everyday Life: The Limits of the Possible* (vol. 1 of *Civilization and Capitalism, 15th–18th Century*). Trans. Sian Reynolds. New York: Harper & Row, 1979.

Cooper, James Fenimore. *The Oak Openings, or the Bee-Hunter.* Boston: Houghton Mifflin, 1881.

Cronon, William. *Nature's Metropolis: Chicago and the Great West.* New York: W. W. Norton, 1991.

Ellul, Jacques. "The Technological Order." *Technology and Culture* 3 (1962), pp. 417ff.

———. *The Technological Society.* Trans. John Wilkinson. New York: Knopf, 1964.

Florman, Samuel C. *The Existential Pleasures of Engineering.* New York: St. Martin's Press, 1976.

Garraty, John A. *The Great Depression.* New York: Doubleday, 1987.

Halberstam, David. *The Reckoning.* New York: William Morrow, 1986.

Hawke, David Freeman. *Nuts and Bolts of the Past: A History of American Technology, 1776–1860.* New York: Harper & Row, 1988.

Hobhouse, Christopher. *1851 and the Crystal Palace.* New York: Dutton, 1937.

Hobsbawm, E. J. *The Age of Revolution, 1789–1848.* New York: New American Library, 1962.

Hughes, Thomas P. *American Genesis: A Century of Invention and Technological Enthusiasm 1870–1970.* New York: Penguin, 1989.

Hurd, Edith Thacher. *Sailers, Whalers and Steamers: Ships That Opened the West.* Menlo Park, Calif.: Lane Book Co., 1964.

Jackson, Donald P. *Gold Dust.* New York: Knopf, 1980.

Jewkes, John, David Sawyers, and Richard Stillerman. *The Sources of Invention.* Second ed. New York: W. W. Norton, 1969.

Koch, Adrienne and William Peden, eds. *The Life and Selected Writings of Thomas Jefferson.* New York: Modern Library, 1944.

Madson, John. *Where the Sky Began: Land of the Tallgrass Prairie.* San Francisco: Sierra Club, 1982.

Marshall, S. L. A. *World War I.* Boston: Houghton Mifflin, 1964.

McNeill, William H. *The Rise of the West: A History of the Human Community.* Chicago: University of Chicago Press, 1991.

McPherson, James M. *Battle Cry of Freedom: The Civil War Era.* New York: Ballantine, 1988.

The Midwest: A Collection from Harper's Magazine [facsimiles of original turn-of-the-century articles]. New York: Gallery Books, 1991.

Mumford, Lewis. *Technics and Civilization.* New York: Harcourt Brace Jovanovich, 1934.

Rolle, Andrew F. *California: A History.* Second ed. New York: Crowell, 1969.

Schlesinger, Arthur M., Jr. *The Age of Jackson.* Boston: Little, Brown, 1953.

———. *The Coming of the New Deal.* Vol. 2 of *The Age of Roosevelt.* Boston: Houghton Mifflin, 1959.

Schmidt, K. M. and Michelle R. Schmidt. "Why Inventors and Entrepreneurs War." *Entrepreneur,* May 1985, pp. 39ff.

Smith, Page. *The Nation Comes of Age: A People's History of the Antebellum Years,* vol. 4. New York: McGraw-Hill, 1981.

"Transportation of Grain in the United States." *Scientific American,* October 24, 1891, cover, pp. 258ff.

Trollope, Anthony. *North America.* Donald Smalley and Bradford A. Booth, eds. New York: Knopf, 1951.

Tuchman, Barbara W. *The Proud Tower: A Portrait of the World Before the War 1890–1914.* New York: Bantam Books, 1967.

Vaughan, Floyd L. *The United States Patent System: Legal and Economic Conflicts in American Patent History.* Norman: University of Oklahoma Press, 1956.

Zinn, Howard. *A People's History of the United States.* New York: Harper & Row, 1980.

VII Sources on miscellaneous agrarian subjects

Bailey, Charles W. *The Land Was Ours: A Novel of the Great Plains.* New York: HarperCollins, 1991.

Berry, Wendell. *What Are People For?* San Francisco: North Point Press, 1990.

Berry, Wendell. *The Unsettling of America: Culture and Agriculture.* San Francisco: Sierra Club Books, 1986 (first published 1977).

Clark, John G. *The Grain Trade in the Old Northwest.* Urbana: University of Illinois Press, 1966.

Crevecoeur, J. Hector St. John de. *Letters from an American Farmer and Sketches of Eighteenth-Century America.* Albert E. Stone, ed. New York: Penguin, 1981.

Critchfield, Richard. *Trees, Why Do You Wait? America's Changing Rural Culture.* Washington, D.C.: Island Press, 1991.

Council for Agricultural Science and Technology. *'Alternative Agriculture': Scientists' Review.* Special Report no. 16. Ames: Iowa State University Press, July 1990.

Davidson, Osha Gray. *Broken Heartland: The Rise of America's Rural Ghetto.* New York: Free Press, 1990.

Duvick, Donald N. "What Vision Should We Pursue in the Agricultural Sciences?" Unpublished draft, dated September 30, 1991.

Fitzgerald, Deborah. "Beyond Tractors: The History of Technology in American Agriculture." *Technology and Culture,* vol. 32, no. 1 (Jan. 1991), pp. 114–26.

Grant, H. Roger and L. Edward Purcell. *Years of Struggle: The Farm Diary of Elmer G. Powers, 1931–1936.* Ames: Iowa State University Press, 1976.

Green, Constance. *Eli Whitney and the Birth of American Technology.* Boston: Little, Brown, 1956.

Heady, Earl O., Edwin O. Haroldsen, Leo V. Mayer, and Luther G. Tweeten. *Roots of the Farm Problem: Changing Technology, Changing Capital Use, Changing Labor Needs.* Ames: ISU Press, 1965.

Iowa State College School of Agriculture. *The Midwest Farm Handbook.* Third ed. Ames: Iowa State College Press, 1954.

Jackson, Wes. *Altars of Unhewn Stone: Science and the Earth.* San Francisco: North Point Press, 1987.

———. *New Roots for Agriculture.* New ed. Lincoln: University of Nebraska Press, 1980.

———, Wendell Berry, and Bruce Colman, eds. *Meeting the Expectations of the Land: Essays in Sustainable Agriculture and Stewardship.* San Francisco: North Point Press, 1984.

Kahn, E. J., Jr. *The Staffs of Life.* Boston: Little, Brown 1984.

Klinkenborg, Verlyn. *Making Hay.* New York: Nick Lyons Books, 1986.

Kramer, Mark. *Three Farms: Making Milk, Meat, and Money from the American Soil.* Harvard University Press, 1987.

Mirsky, Jeanette and Allan Nevins. *The World of Eli Whitney.* New York: Macmillan, 1952.

Morgan, Dan. *Merchants of Grain.* New York: Penguin, 1980.

Nabhan, Gary Paul. *Enduring Seeds: Native American Agriculture and Wild Plant Conservation.* San Francisco: North Point Press, 1989.

National Research Council. *Alternative Agriculture.* Washington, D.C.: National Academy Press, 1989.

Paul, Rodman W. "The Great California Grain War: The Grangers Challenge the Wheat King." *Pacific Historical Review,* XXVII no. 4 (Nov. 1958), pp. 331–49.

———. "The Wheat Trade Between California and the United Kingdom." *Mississippi Valley Historical Review,* XLV no. 3 (Dec. 1958), pp. 391–412.

Pollan, Michael. *Second Nature: A Gardener's Education.* New York: *Atlantic Monthly,* 1991.

Rasmussen, Wayne D. *Taking the University to the People: Seventy-five Years of Cooperative Extension.* Ames: Iowa State University Press, 1989.

Reisner, Marc. *Cadillac Desert: The American West and Its Disappearing Water.* New York: Penguin, 1986.

Rhodes, Richard. *Farm: A Year in the Life of an American Farmer.* New York: Simon & Schuster, 1989.

Rothstein, Morton. "Antebellum Wheat and Cotton Exports." *Agricultural History,* XL no. 2 (Apr. 1966), pp. 91–100.

Soth, Lauren. *An Embarrassment of Plenty: Agriculture in Affluent America.* New York: Thomas Y. Crowell, 1965.

Strange, Marty. *Family Farming: A New Economic Vision.* Lincoln: University of Nebraska Press, 1988.

Taylor, Henry C. *Tarpleywick: A Century of Iowa Farming.* Ames: Iowa State University Press, 1970.

U.S. Department of Agriculture. Milton S. Eisenhower, ed. *Yearbook of Agriculture, 1933.* Washington: Government Printing Office, 1933.

———. *Farmers in a Changing World: Yearbook of Agriculture, 1940*. Washington: Government Printing Office, 1940.

———. *Science in Farming: Yearbook of Agriculture, 1943–1947*. Washington: Government Printing Office, 1947.

———. *Power to Produce: Yearbook of Agriculture, 1960*. Washington: Government Printing Office, 1960.

———. *After a Hundred Years: Yearbook of Agriculture, 1962*. Washington: Government Printing Office, 1962.

———. *Protecting Our Food: Yearbook of Agriculture, 1966*. Washington: Government Printing Office, 1966.

INDEX

Page numbers in *italics* indicate illustrations.